Climate Hazard Crises in Asian Societies and Environments

Climate hazards are the world's most widespread, deadliest and costliest natural disasters. Knowledge of climate hazard dynamics is critical since the impacts of climate change, population growth, development projects and migration affect both the impact and severity of disasters. Current global events highlight how hazards can lead to significant financial losses, increased mortality rates and political instability.

This book examines climate hazard crises in contemporary Asia, identifying how hazards from the Middle East through South and Central Asia and China have the power to reshape our globalised world. In an era of changing climates, knowledge of hazard dynamics is essential to mitigating disasters and strengthening livelihoods and societies across Asia. By integrating human exposure to climate factors and disaster episodes, the book explores the environmental forces that drive disasters and their social implications. Focusing on a range of Asian countries, landscapes and themes, the chapters address several scales (province, national, regional), different hazards (drought, flood, temperature, storms, dust), environments (desert, temperate, mountain, coastal) and issues (vulnerability, development, management, politics) to present a diverse, comprehensive evaluation of climate hazards in Asia. This book offers an understanding of the challenges climate hazards present, their critical nature and the effort needed to mitigate climate hazards in 21st-century Asia.

Climate Hazard Crises in Asian Societies and Environments is vital reading for those interested and engaged in Asia's development and well-being today and will be of interest to those working in Geography, Development Studies, Environmental Sciences, Sociology and Political Science.

Troy Sternberg is a researcher at the School of Geography, Oxford University. His research focuses on climate hazard impact on environments and societies across Asian drylands.

Routledge Studies in Hazards, Disaster Risk and Climate Change

Series Editor: Ilan Kelman

Reader in Risk, Resilience and Global Health at the Institute for Risk and Disaster Reduction (IRDR) and the Institute for Global Health (IGH), University College London (UCL)

This series provides a forum for original and vibrant research. It offers contributions from each of these communities as well as innovative titles that examine the links between hazards, disasters and climate change, to bring these schools of thought closer together. This series promotes interdisciplinary scholarly work that is empirically and theoretically informed, with titles reflecting the wealth of research being undertaken in these diverse and exciting fields.

Published

Cultures and Disasters
Understanding Cultural Framings in Disaster Risk Reduction
Edited by Fred Krüger, Greg Bankoff, Terry Cannon, Benedikt Orlowski and E. Lisa F. Schipper

Recovery from Disasters
Ian Davis and David Alexander

Men, Masculinities and Disaster
Edited by Elaine Enarson and Bob Pease

Unravelling the Fukushima Disaster
Edited by Mitsuo Yamakawa and Daisaku Yamamoto

Rebuilding Fukushima
Edited by Mitsuo Yamakawa and Daisaku Yamamoto

Climate Hazard Crises in Asian Societies and Environments
Edited by Troy Sternberg

Climate Hazard Crises in Asian Societies and Environments

Edited by
Troy Sternberg

Routledge
Taylor & Francis Group
LONDON AND NEW YORK

First published 2017
by Routledge
2 Park Square, Milton Park, Abingdon, Oxon OX14 4RN

and by Routledge
711 Third Avenue, New York, NY 10017

Routledge is an imprint of the Taylor & Francis Group, an informa business

British Library Cataloguing-in-Publication Data
A catalogue record for this book is available from the British Library

Library of Congress Cataloging-in-Publication Data
A catalog record for this book has been requested

ISBN: 978-1-4724-4646-6 (hbk)
ISBN: 978-1-315-57241-3 (ebk)

Typeset in Times New Roman
by Apex CoVantage, LLC

Contents

Figures

Tables

Contributors

Madina Batyrbaeva and **Irina Vitkovskaya** are leading scientists at the Department of Earth Monitoring of the National Centre of Space Research and Technology of Kazakhstan. They deal with collecting and processing of the daily remote sensing information for the study of land degradation in Kazakhstan, droughts monitoring, vegetation productivity and fire detection.

Yung-Fang Chen's research is on the pedagogy of emergency response training and exercises. Her current projects and interests include tasks for post-disaster assistance, shelter and housing, community reconstruction, e-learning and risk communication.

Tim Colbourn is a Lecturer in Global Health Epidemiology and Evaluation at University College London, England.

Lena Dominelli, Co-Director of the Institute of Hazard, Risk and Resilience at Durham University, has undertaken research projects on climate change, flooding, earthquakes and volcanoes. She is a leading figure in social work and chairs the Disaster Interventions Committee for the International Association of Schools of Social Work, and through it represents social work at UNFCCC COP21 meetings.

Francesco Femia is Co-Founder and President of the Center for Climate and Security, where he co-leads the Center's policy development, analysis and research programmes.

Rodica Indoitu and **Giorgi Kozhoridze** are PhD students at J. Blaustein Institutes for Desert Research. Rodica's PhD thesis deals with temporal and spatial variations of dust storms in the Middle Asia, while Giorgi studied the land cover changes occured on the dried bottom of the Aral Sea.

Ilan Kelman is a Reader in Risk, Resilience and Global Health at University College London, England and a researcher at the University of Agder, Kristiansand, Norway.

Helen Lackner is an independent researcher with over 40 years' experience in Yemen and is writing an introduction to the crisis in Yemen to be published in 2017. She worked as a rural development consultant in over 30 countries.

Batyr Mamedov is from the National Institute of the Deserts, Flora and Fauna, Turkmenistan. His research interests include water management and soil conservation for both agricultural and environmental objectives; and development-oriented research on desertification and the effects of climate change.

Leah Orlovsky is a Senior Researcher at the Swiss Institute for Dryland Environmental & Energy Research, J. Blaustein Institutes for Desert Research of Ben-Gurion University, Israel. Her scientific interests include the use of remote sensing for detecting changes in land cover and land use, and analysing the impacts of these changes on climate, environment and society.

Nikolai Orlovsky from J. Blaustein Institutes for Desert Research, Israel is a climatologist and expert in desertification problems in Central Asia. His primary interest is the study of the climate change and its infuence on land degradation processes.

Prajjwal Panday is geographer with research and intellectual interests focused on using a systems approach to determine linkages between climate variability, anthropogenic changes and land-water interactions. His research focuses on how climate variability and land use change influence water balances in Himalayan and Amazonian ecosystems.

Inam-ur-Rahim is director of the Foundation for Research and Socio-Ecological Harmony, Islamabad, Pakistan and freelance researcher on livestock systems and climatic and environmental issues.

Henri Rueff is a Geographer interested in smallholder's livelihoods living in resource scarce and remote areas in mountains and deserts. He works at the Department of Environmental Sciences, University of Basel, Switzerland.

Masato Shinoda, DSc is a Professor of the Graduate School of Environmental Studies at Nagoya University in Japan. His research field is ecological climatology of drylands.

Troy Sternberg is a researcher at the School of Geography, Oxford University. His research focuses on climate hazard impact on environments and societies across Asian drylands.

Caitlin Werrell is Co-Founder and President of the Center for Climate and Security, where she co-leads the Center's policy development, analysis and research programmes.

Foreword

Climate hazards are the world's most universal and damaging natural disasters. Floods, droughts and storms affect tens of millions of people, disrupt livelihoods, negatively impact communities and impoverish residents. At the same time these events are harmful to societies, economics, infrastructure and environments. Our book, *Climate Hazard Crises in Asian Societies and Environments*, examines the multiple dimensions of climate hazards across Asia. Exploring the social implications and environmental dynamics of risk, mitigation and vulnerability are a major global challenge. The book's contemporary focus highlights the challenges hazards present in Asia as climate change, expanding populations, rapid development and conflict affect the continent's 4.4 billion people. The chapters cover a diverse range of climate hazard contexts that are essential reading for understanding the ongoing threat climate hazards present for Asia in the 21st century.

Natural hazards become disasters when they affect people and society. In Asia climate disasters now account for 87 per cent of hazard events and >90 per cent of fatalities. Their frequency, magnitude and intensity affects the basics of life – water, food, health and livelihoods. The book's key theme is how recurrent disasters are shaped by people and landscapes as well as climate. Trigger events reflect a combination of weather forces, social exposure and the ability to mitigate disaster impact and physical damage. The book crosses the continent to identify the many forms hazard risks take in Asia. Beginning in the Middle East and Arabia, the chapters address key issues of drought and environments, then moves on to South Asia to highlight complex flood dynamics in Pakistan and the impact of global warming in Himalayas. Dust in Central Asia reflects the trans-border nature of hazard events as well as how policy decisions can dramatically increase hazard exposure. In East Asia Taiwan offers a well organised storm response strategy whilst China exemplifies the natural vs constructed risk conundrum hazards may present. Large-scale assessments integrate the weather and disaster management over Eurasia, dryland exposure across the continent, the hazard implications on health and the effects of human and political action on societal resilience. The range and breadth covered convey the complexity of climate hazards whilst providing critical insight into hazard immediacy at local and regional levels.

Commencing with Femia and Werrell's interpretation of climate and the Arab Spring provides a cautionary tale of how drought, poor governance and migration

affected social dynamics in Syria; research identifies the role of climate hazards in instigating civil war. When writing on Yemen, Lackner had to twice update her chapter on climate, development and security in Yemen due first to political conflict, then civil war fueled by international forces. In both countries, floods, drought or environmental stress contributed to conflict yet are forgotten amidst state collapse. Rahim and Rueff capture the multiple factors that make Indus floods so deadly in Pakistan. Inequality, land tenure rights, weak governance and economic exploitation ignore endemic flooding at great human and social cost. Panday examines how changing global weather patterns in the Hindu Kush-Himalayan region affect water resources, ecosystems and livelihoods for >600 million downstream agriculturalists.

More positively, Chen traces Typhoon Morakot 88 and the resulting flood disaster's impact in Taiwan. Whilst to many the government response was exemplary, particularly compared to global norms (e.g. Hurricane Katrina in the US), satisfying local concerns proved elusive. Sternberg's evaluation (Chapter 8, this volume) of underlying forces framing China's hazard regime makes clear that humans and governments can contribute to, as well as mitigate, climate disasters. Manipulating nature and the environment raises the question across China whether disaster results from natural forces, policy and human action or a deadly combination of both.

Turning the focus to weather Orlovsky et al.'s long-term investigation of dust storms in the Aral Sea and former Soviet Middle Asia highlights how the severity of extreme dust storms was linked to the human-driven desiccation of the Aral Sea. Investigation finds that since 1990 dust events and hazard damage has decreased. High-impact weather is stressed by Shinoda in the mid-latitude drylands that cover much of the Asian steppe. Efforts to integrate major hazards show the links between drought, *dzud* (extreme cold), dust and desertification which can reshape livelihoods and environments. Research represents a new methodology to identify extreme weather and develop an effective early warning system. Sternberg then expands this theme to present climate hazards in Asian drylands, identifying that arid and semi-arid areas, comprising 39 per cent of Asia, share several similarities when facing climate hazards. Vast spatio-temporal scales, drought, marginality and policy neglect combine to increase exposure. Improved awareness and international cooperation can strengthen knowledge and reduce vulnerability across desert societies.

More broadly, Kelman and Colbourn bring the issue of hazard and health in Asia to the forefront. This is a cross-cutting issue from tropics to mountains that reflects poverty and location-borne exposure that varies in disease and incidences as well as corollary impacts on food and water safety. Interactions include human physical, environmental and sociological health. Dominelli draws several themes together in her chapter on society and disaster. She stresses the human role in disaster engagement, gaining political support for action, then integrating the local and the global when working across boundaries and borders. The great social impact of hazards shows the need for new theory and practice that emphasises resilience throughout the disaster process.

The chapters show the many ways that climate hazards impact and interact with societies. Each environment has recognised exposure to climate events whilst

human and social pressures add a level of risk and vulnerability for residents in Asian landscapes. We are often aware of physical dimensions but pay less attention to socio-economic dynamics like marginality, education and politics. In addition, capacity and funding for effective mitigation of hazards varies across countries. Integrating the natural, social and governance roles can identify how disasters may be directly linked to specific risk factors and how system failure can turn climate events into crises. Disasters are multi-faceted events best mitigated through a holistic approach that encompasses the range of contributory factors in each country and landscape.

Chapters reflect the great diversity of Asian environments and the relevant hazard dynamics. Deserts are represented in work on Syria, Yemen and Central Asia whilst the two dryland chapters stress common hazard patterns and exposure that are specific to arid zones. As an example, Pakistan shows how agricultural heartlands can also be part of extensive drylands nourished by water from mountain sources. The same water that brings fecundity can also wreak great damage through flooding. The source for the Indus in Pakistan and Ganges in India are in the Himalayas, making changes to mountain climates a key hazard risk. Of particular concern in a warming world are rates of glacial melt which directly affect water resources for hundreds of millions of farmers in South Asia. The greater Himalayan alpine region also provides the source water for the Yangtse, Yellow, Mekong and Brahmaputra rivers. Typhoon Morakot in Taiwan and Typhoon Haiyan in the Philippines identify how coastal regions are particularly vulnerable to storm activity and resulting community exposure.

The multiple perspectives covered in the book explore the socio-economic centrality of hazard mitigation. Perhaps it should be no surprise that Taiwan, with the highest GDP of countries represented herein, has provided the most effective example of hazard mitigation in the case studies. The corollary finds impoverished states, such as Yemen and Pakistan, vulnerable to climate forces that exacerbate endemic social challenges. China provides another approach to hazard management that responds to natural events whilst also exacerbating local exposure. This suggests that over-management can be detrimental as poor management is damaging. Appraisal chapters show how, on diverse issues from health to exposure to drylands, mitigation capacity depends on several inter-linked factors. For instance, drawing together landscapes and human action, we see that dust in Central Asia was directly linked to centralised economic planning that encouraged agriculture despite great downstream ecological damage that turned a flood region into a major dust source.

The nuanced role of climate hazards is often missed amongst dramatic event headlines, great human misery (migration, conflict, mortality), timespans and spatial scales as well as limited coverage or interest amongst international audiences to localised events. At the same time the immediacy of hazards focuses engagement on initiation and damage but seldom takes a longer-term perspective to understand the forces that contribute to disaster, or an assessment-over-time of how hazards impact society. A clear example is how the Hindu Kush-Himalayan chapter directly links changes in global warming with the mountain climate and

downstream agricultural systems. Thus as a warming temperature affects glacial melt, water supply and flood risk that threatens the future viability of farming in the Gangetic and Indus plains. Chapters on Syria and Yemen work backwards to show how past climate events reduced capacity to cope with crises. The resulting conflicts show how poor disaster management contributes to great destruction. Similarly, the conceivable risk of the Three Gorges Dam being breached in a flood would result in a catastrophic event.

The many threads of hazard implications place responsibility for mitigation strategy on national and local governments. Political leadership, effective policy and local action are key across Asia yet chapters reflect a reluctance of the state to engage and address root-and-stem causes and issues. The cost, time and attention may be one reason; another is the difficulty in bringing together multiple interests and perspectives to resolve challenges. In countries with low capacity, the international community often plays an additional role. Effective preparation is repeatedly stressed in global forums, such as the UN International Strategy for Disaster Reduction and the Hyogo Framework for Action 2005–2015. These well-intentioned approaches and documents offer guidelines to improve disaster engagement. However, the fact remains that countries, especially in the developing world, face several immediate challenges, such as food, health, education and security that take precedence over unpredictable disasters that may later become exponentially more damaging. We see this book as call to action for Asian leaders, researchers and civil society because neglecting climate disasters changes hazards into crises events. The first step is acknowledgement and awareness by all stakeholders. Then, effective methods to engage with disasters are needed. Finally, adequate mitigation mechanisms are essential to limiting damage. Avoiding hazards is unrealistic; however, basic response systems can keep disasters from spinning out control. Poor reaction to a disaster exacerbates a host of vulnerabilities and triggers a cascading set of further problems and misery. Throughout hazard evaluation we need to keep sight of their human costs in lives lost, livelihoods disrupted, homes and work destroyed and hope extinguished.

Often politicians and publics engage with research at a distance, yet in-depth investigation is key to identifying problems, breakpoints, systemic weaknesses and response failures. More positively, clear knowledge provides an excellent starting point for addressing hazards. Understanding climate variability and change; household, community and national exposure and potential mitigation strategies at several levels and learning from the experience of others offer ways to enhance and restructure climate hazard thinking, planning and response. The chapters in this book give insight into climate hazards across Asia. In such a vast continent, this work is an introduction and provides a way to grasp and understand the complexity, intensity and immediacy of hazards in our globalised world. In each chapter, we see the role of humans and their societies. Our challenge is to take this a step further – that once we better understand hazards to then engage and address hazard risk. Whilst the challenge is global, *Climate Hazard Crises in Asian Societies and Environments* provides an introduction to climate hazard crises in Asia.

1 An unstable, stable nation? Climate, water, migration and security in Syria from 2006–2011

Francesco Femia and Caitlin Werrell

Introduction

From 2006–2012 Syria experienced one of the worst extended droughts in its history. This drought, coupled with natural resource mismanagement, demographic dynamics and overgrazing in certain areas, contributed to a massive displacement of agricultural and pastoral peoples. Despite these dynamics and other existing underlying socio-political and economic grievances, key actors in the international community largely considered Syria to be a stable country relative to other nations in the Middle East and North Africa that experienced significant social unrest in the so-called Arab Uprisings (Butters 2011; Mann 2012). This chapter explores the climate and natural resource elements of Syria's state fragility, the phenomenon of governments across the international community misdiagnosing the probability of Syrian instability in 2011 and the possible pathways forward for the country and the international community in addressing these risks.

The climate-water-natural resource management nexus in Syria from 2006–2012

The factors that contributed to the popular uprisings in Syria in 2011 are very complex and remain little explored. As with all conflicts, a confluence of ultimate and proximate causal factors intersect, resulting in discontent turning to revolt, and governments either managing or suppressing that revolt, collapsing or something in between. In the case of Syria's popular revolt, which began most visibly in the southern rural town of Dara'a in March 2011 (PBS 2011), political, economic, ethnic, sectarian and religious grievances, as well as inspiration from uprisings in Tunisia and Egypt, have been offered as contributing factors to the collapse of security in the country.

Less attention has been paid, however, to significant agricultural, pastoral, environmental and climatic changes in Syria. Combined with the mismanagement of water and food resources by the al-Assad regime, which between 2007 and 2011, these changes converged to precipitate a severe humanitarian crisis. Despite UN reports highlighting the crisis, it was barely noticed by the international community, in part due to the Syrian government's attempts to prevent reporters from

accessing internally displaced peoples (Worth 2010). This study represents an update of a previous study on the subject (Femia and Werrell 2012).

A climate-exacerbated drought on top of a drought

From 2006 to 2012, Syria experienced one of the worst long-term droughts and most severe set of crop failures and livestock devastation in its history of records, with the period from 2009–2012 registering as the most extreme drought conditions across a number of regions (Werrell et al. 2015, p. 31). From 2007 to 2008, the severe drought affected 97.1 per cent of Syria's vegetation (Figure 1.1) (Wadid et al. 2011, p. 11). This drought also followed on the heels of another of Syria's most severe droughts in modern history, which took place from 1999–2000, and affected 329,000 people (Werrell et al. 2015, p. 31).

Recent evidence suggests that the probability of such a severe-to-extreme drought period from 2006–2012 increased as a result of anthropogenic climate change. A study by Hoerling et al. (2012) found strong evidence that winter precipitation decline in the Mediterranean littoral and the Middle East from 1971 to 2010 was likely due to climate change, with the region experiencing nearly all of its driest winters since 1902 in the past 20 years (Figure 1.2) – a problematic phenomenon given that the region receives most of its annual rainfall in the

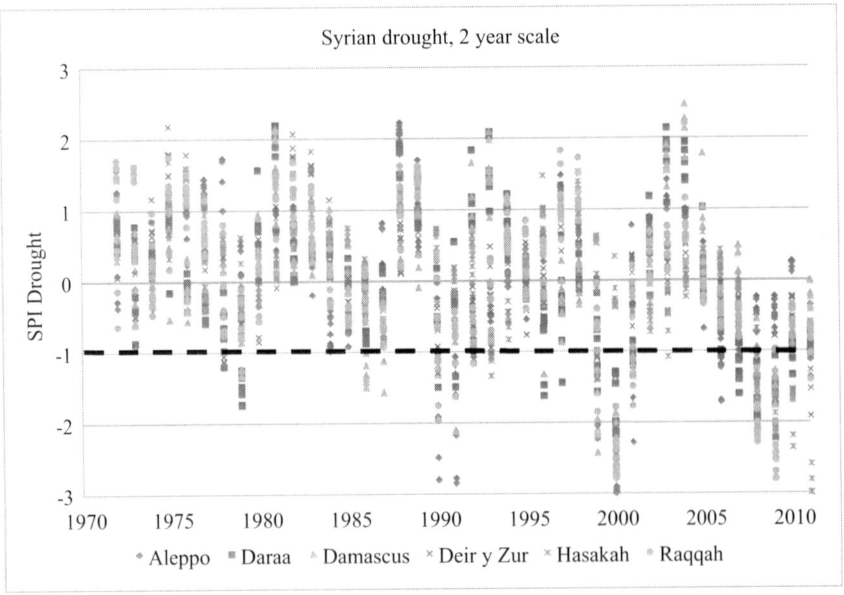

Figure 1.1 Drought, two-year scale, Syria from 1970 to 2011. Calculated by Standard Precipitation Index (SPI), -1 or below signifies drought (-1 = moderate drought; -2 = extreme drought)

Source: Sternberg, unpublished data. For SPI calculation see Sternberg (2012)

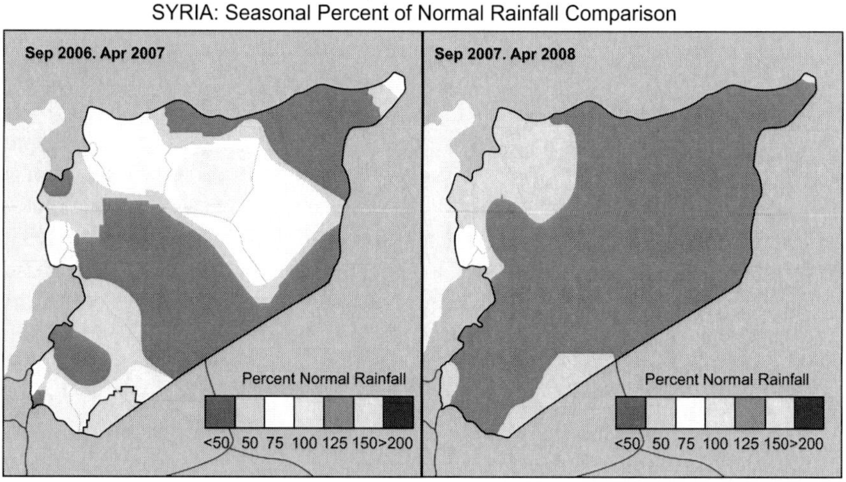

Figure 1.2 Syria: Seasonal per cent of normal rainfall, comparison 2006–2008 (USDA 2008)

Source: USDA Foreign Agricultural Service (FAS. 2008. SYRIA: Wheat production in 2008/09 declines owing to season-long drought http://www.fas.usda.gov/highlights/2008/05/syria_may2008.htm)

winter. This trend of precipitation decline can also be seen quite clearly in the Standard Precipitation Index (Werrell et al. 2015, p. 31). The authors determined that half of this drying magnitude can be explained by anthropogenic greenhouse gas and aerosol forcing, as well as increases in sea surface temperature (Hoerling et al. 2012).

More recently, a study by Kelley et al. (2015) found that the extreme drought in Syria from (2007–2010) was two to three times more likely to be a result of anthropogenic rather than natural climatic changes (Kelley et al. 2015).

Natural resource mismanagement and desertification

The reasons behind the collapse of Syria's farmland and rangeland extend beyond the drought, and the climate change drivers that increased its likelihood. A complex interplay of variables, including natural resource mismanagement, demographic dynamics and overgrazing interacted with changing climatic conditions to enable that outcome (Femia and Werrell 2012).

First, poor governance by the al-Assad regime compounded the effects of the drought, which contributed to water shortages and land desertification. The al-Assad government, like many of its predecessors, heavily subsidised water-intensive wheat and cotton farming (more than 50 per cent of which was grown in the al-Hassakeh governorate (Matlock 2008), which was incidentally hardest hit by the drought). A focus on water-intensive crops, coupled with the widespread use of inefficient irrigation techniques, such as 'flood irrigation,' wherein nearly 60 per

cent of water used is wasted, placed significant strains on Syria's water resources (IRIN 2010a). This dynamic stood in contrast to a number of other nations in the Middle East and North Africa, who, for example, imported most of their wheat. In fact, 9 out of the top 10 wheat importers in the world can be found in the Middle East and North Africa (Sternberg 2013, p. 13).

In the face of water shortages that flowed from water-intensive agricultural practices, the previous drought from 1999–2000 and population pressures, farmers sought to increase supply by turning to the country's groundwater resources. Syria's National Agricultural Policy Center reported a 63 per cent increase in wells tapping aquifers from 1999 to 2007 (Sticklor 2010). This pumping 'caused groundwater levels to plummet in many parts of the country, and raised significant concerns about the water quality in remaining groundwater stocks'(Sticklor 2010). This wheat and cotton production, coupled with the severe–extreme drought, significantly diminished the water table in the country.

On top of water resource mismanagement, the Food and Agriculture Organization (FAO) reported that the overgrazing of land, and a rapidly growing population, compounded the land desertification process in Syria (IRIN 2010b). In a study from 2014, De Châtel (2014) also determined that overgrazing in areas of Syria affected by drought may have been a key driver of desertification (De Châtel 2014).

Water resource management in neighbouring countries, particularly Turkey, may also have played a role in Syria's water insecurity during the decade prior to the uprising, though likely not as significant a role as the al-Assad government's own policies. In particular, from 1990–2010, average annual flows of the Euphrates River, as measured close to the Turkish border in Jarabulus, were significantly lower than the average annual flows from 1937–1990. This decline coincided with the 1992 completion of the Ataturk Dam on the Euphrates in southeastern Turkey (Gleick 2014, p. 5).

Internal mass displacement

This climate-drought-natural resource management nexus in Syria ultimately precipitated a significant and under-reported humanitarian disaster in the country from 2006–2011, prior to the outbreak of widespread popular revolt against the al-Assad government.

Of the most vulnerable Syrians dependent on agriculture, particularly in the northeast governorate of Hasakah, 75 per cent experienced total crop failure (Gleick 2014, p. 15). On average, pastoral peoples in the northeast of the country lost around 85 per cent of their livestock (Worth 2010). As of 2010, the combined impact on agricultural and pastoral lands affected at least 1.3 million people (Werrell et al. 2015, p. 32).

According to a report from *the New York Times*, the al-Assad government attempted to prevent international observers and journalists from accessing people affected by the collapse of farmland and rangeland (EM-DAT 2016). Nonetheless, from 2009 through 2011, some institutions of international agencies and non-governmental

organisations had begun to identify the humanitarian crisis unfolding in Syria. In 2009, both the United Nations and International Red Cross reported that over 800,000 Syrians had lost their livelihoods as a result of the agricultural and rangeland collapse (IRIN 2009). By 2011, it was estimated that one million Syrians had been left extremely 'food insecure' by the collapse (Wadid et al. 2011, p. 5). Another estimated that two to three million people had been driven into extreme poverty, approximately 9–13 per cent of the country's population (Worth 2010).

This loss of livelihoods led directly to a mass displacement of farmers, herders and agriculturally-dependent rural families, a majority of whom moved to urban areas in search of employment opportunities (Wadid et al. 2011, p. 8). One estimate suggested that between 1.5 and 2 million people had been displaced (Mohtadi 2012). In October 2010, the a UN estimate that 50,000 families migrated from rural areas just that year, following the hundreds of thousands of people that had migrated to urban areas during the previous years of the drought (Worth 2010). In January 2011, it was reported that crop failures just in the farming villages around the city of Aleppo drove roughly '200,000 people from rural communities into the cities' (Nabhan 2010). This occurred while Syrian cities were already coping with influxes of Iraqi refugees since 2003, as well as a steady stream of refugees from Palestinian territory (UNHCR 2010). Crumbling urban infrastructure, a phenomenon that preceded the drought, combined with these population pressures and contributed to a significant decline in per capita water availability in urban areas (IRIN 2010a).

Interaction of natural resource stress and socio-political dynamics

The stresses that flowed from drought conditions, as well as water and land mismanagement by the al-Assad regime, existed in the context of a range of socio-political grievances among non-Alawite Arab and Kurdish populations in rural areas of Syria, particularly in the north and south. For example, anecdotal evidence suggests that well-drilling contracts were often awarded by the al-Assad government on sectarian grounds, favouring Alawite and other Shiite populations over others ("Anonymous" 2013).

The rural farming town of, the focal point for protests in the early stages of the opposition movement in 2011, was home to all of the stresses detailed above. Dara'a had been significantly affected by five years of drought and water scarcity (PBS 2011). The town also hosted a population that had been largely ignored by the al-Assad government, not least due to sectarian differences (Paralleli 2011).

Recent research has also made the case that discontent among a number of tribal populations displaced by the drought, a dramatically declining water table and underdevelopment played an important role in tribal uprisings, despite attention being paid primarily to sectarian drivers of unrest (Dukhan 2014). According to Syrian researcher Haian Dukhan:

> the collapse of the rural economy of tribal communities in the south and east of Syria during Bashar al-Assad's regime due to drought, lack of development

> projects and the mismanagement of al-Badia resources ignited the Syrian uprising to start in tribal regions.

Lastly, anonymous, unpublished interviews with tribal peoples who were displaced by the drought, and migrated to suburbs of Damascus and Homs, describe evidence of social tensions. This includes tensions between tribespersons from Hassakeh (particularly the Jabbour and Tay tribes) and residents of the Hajr Aswad suburb of Damascus, as well as sectarian tensions between displaced members of the Fwaira and Nu'im tribes and Alawite residents of the Baba Amr suburb in Homs (Dukhan 2015).

Though the lines of causality remain difficult to disentangle, it is reasonable to suggest that the combination of climate, drought, natural resource mismanagement, internal mass displacement of farmers and herders and sectarian grievances in Syria, coupled with knowledge of the recent revolutions in Tunisia and Egypt, played a role in fraying the social contract between a range of rural and urban populations in the country and the seemingly 'stable' al-Assad government (Shadid 2011).

Misdiagnosed: Syria's unstable stability

Despite the climate, water and food insecurities in Syria detailed earlier, and despite the mass displacement of people that followed, key actors in the international community, including the US government, seemed largely surprised by the Syrian uprisings that began in 2011. For example, the US Deputy Secretary of State during the initial wave of the Arab Uprisings, James Steinberg, was very clear about the fact that no one in a position of authority within the US foreign policy infrastructure considered Syria to be a likely candidate for significant political unrest. In an interview, Steinberg highlights the fact that Syria sat at the bottom of an Administration list of Middle Eastern and North African nations that were 'at risk of large-scale political turmoil'(Mann 2012, p. 270). Given that such a list was likely compiled using analysis from key departments and agencies in the US government, including the intelligence community, it can be reasonably surmised that predictive tools being utilised at the time by US government analysts were missing some important elements:

> The Obama administration began trying to figure out which country would come next. Administration officials hurriedly made a list of which countries in the Middle East were most at risk of large-scale political turmoil, and which were least at risk. That list turned out to be wrong in many cases. At the top of the list were Yemen and Jordan, the countries where political unrest seemed likely. In the middle tier were Libya, Bahrain and Oman, all countries where it appeared possible. At the bottom were the nations where any widespread demonstrations for democracy were judged to be improbable: Saudi Arabia and Syria. 'No one was focused on Syria, because it seemed far less likely than other states in the region,' said James Steinberg a few months later.
>
> (Mann 2012, p. 270)

Analysts in the public sphere also seemed to have not seen the Syrian uprising coming. During the progress of the uprisings in Tunisia and Egypt, few (if any) reporters in English language media suggested a high probability of a Syrian uprising that could threaten the al-Assad regime. On Feb 4, 2011, the eve of political turmoil in the country, *TIME* magazine's Andrew Lee Butters, who had recently witnessed political protests in Damascus, determined that a government's alliance with the United States was a key indicator of a fragile government, and that Syria did not meet that criteria. As he stated, less than a month before the Syrian government affected a violent clampdown of political unrest in Dara'a:

> demonstrations in Syria are unlikely to pick up anywhere near enough momentum to seriously threaten the regime of President Bashar al-Assad. The reason is simple: Syria, unlike Egypt, Yemen and Jordan, isn't allied with the United States.
>
> (Butters 2011)

Publicly-available measurements of state fragility and climate vulnerability also seemed to have missed deteriorating social, environmental and natural resource conditions in Syria during the five years prior to the uprisings. A study published Werrell et al. (2015) demonstrated that two popular indices used to measure state fragility and climate vulnerability (the Fragile States Index and the Notre Dame Global Adaptation Index, respectively), both seemed to miss a deteriorating trend in the nation's fragility and climate vulnerability (Werrell et al. 2015). In fact, the indices detected an improving trend in both measurements from 2006–2010 (Werrell et al. 2015, pp. 38–39). It is not until 2011, after Syria had already plunged into large-scale political unrest, that the indices detected a deteriorating trend (Werrell et al. 2015, pp. 38–39).

Furthermore, intelligence analyst Margolis (2012) found that from 2007–2010, four respected indices measuring state instability (the Failed States Index, the State Fragility Index, the Economist Intelligence Unit and the Index of State Weakness) presented a picture of improving stability in the country, with Syria ranking as more stable than 48, 75, 94 and 59 other countries in the world, respectively (Margolis 2012).

Additionally, a number of confidential sources within the US government noted that after the uprisings in Tunisia and Egypt, a majority of analysts from across the US intelligence, foreign policy and defense communities had determined that the al-Assad regime was very stable relative to most other nations in the region, and that water insecurity, food insecurity and population displacement dynamics on the sub-national level were not deemed significant enough to increase the likelihood of large-scale political revolt. This latter assessment of the relatively low weight given to natural resource security variables within these predictive assessments was often cited as a 'norm' among regional analysts within the US government. Normally 'accurate' predictive tools not being particularly sensitive to natural resource dynamics were also cited on a number of occasions as one possible, though partial, explanation for this misdiagnosis ('Anonymous' 2013).

In short, evidence suggests that the indices and predictive tools used by political analysts to examine state fragility, and the likelihood of state fragility, may be inadequately incorporating natural resource drivers, and/or may be missing hidden drivers that remain 'unknowns'. In this context, there may be space for researchers and practitioners to develop a broader understanding of how climate and natural resource stresses interact with other known drivers of state fragility and how that relationship can be appropriately reflected in state fragility measurement tools. Far from being an academic exercise, such a process could provide a critical analytical foundation for improving government preparedness for destabilising trends in the future.

The future of climate, water and food security in Syria

While attention remains and should remain on the immediate question of resolving the conflict in Syria, planning for a post-conflict Syria should begin as soon as possible, not least because solutions to water resource questions, in particular, have historically played a role in facilitating cooperation among conflicting parties. Furthermore, climate, water and agriculture prognoses for the region, including Syria, is worrying. If left unresolved, these stresses could continue to present serious challenges to the viability of the Syrian state in the future. Even without the significant deterioration of Syria's water and food infrastructure that has resulted from the conflict (UNICEF 2015), the country is likely to face severe challenges in this area due to climate change-related stresses, including significant precipitation decline and a history of water-intensive crop production, which is likely to resume in a post-conflict environment (Femia and Werrell 2012).

In terms of the country's climate future, the IPCC's 5th Assessment Report forecasts a decrease in winter precipitation in the region as average global temperatures continue to increase (Hijioka et al. 2014). Water security projections are also dire. The 2011 Intelligence Community Assessment on Global Water Security notes that '[b]y 2025, ISciences projects that water stress will increase significantly in many locations throughout the world, including North Africa, the Middle East, and Asia' (DNI 2012).

Lastly, climate projections suggest increased food insecurity in the Syria. For example, models conducted by the International Food Policy Research Institute (IFPRI) in 2011 determined that if current rates of climate change continue, yields of rainfed crops in the country may decline 'between 29 and 57 percent from 2010 to 2050' (Breisinger et al. 2010).

Conclusion

The popular revolt in Syria was treated with surprise by many analysts in the international community, including governments such as the United States for whom stability in the Middle East is a core concern. This seems to be reflected in the statements of senior US government officials, mainstream journalists familiar with the region, anonymous conversations with US intelligence, foreign policy

and defense analysts and popular state fragility indices, such as the Fragile States Index, which surprisingly showed a general trend of increasing stability in Syria over the five years prior to popular uprising.

In this context, the way nations and analysts measure state fragility may need to take into account (or give greater weight to) a broader array of factors, including climate, water and food insecurity. In the absence of a broadening of the definition of state stability, governments and the international community may continue to misdiagnose the level of stability a nation or region enjoys. A more comprehensive understanding of state fragility will likely be an important prerequisite for governments and publics better preventing, preparing for and mitigating destabilising trends.

References

Breisinger, C., van Rheenen, T., Ringler, C., Nin Pratt, A., Minot, N., Aragon, C., . . . Zhu, T. (2010). Food Security and Economic Development in the Middle East and North Africa: Current State and Future Perspectives – 2418.pdf. International Food Policy Research Institute.

Butters, A. L. (2011). Why the Arab Democracy Wave Is Unlikely to Reach Syria – Yet. *Time Magazine*. http://content.time.com/time/specials/packages/article/0,28804,2045328_2045333_2046288,00.html.

De Châtel, F. D. (2014). The role of drought and climate change in the Syrian uprising: Untangling the triggers of the revolution. *Middle Eastern Studies*, 50, 521–535. doi:10.1080/00263206.2013.850076

DNI. (2012). Global Water Security Intelligence Community Assessment. Office of the Director of National Intelligence.

Dukhan, H. (2014). Tribes and tribalism in the Syrian revolution. *Syria Studies Journal*, 6, 1–28.

Dukhan, H. (2015). Anonymous interview.

EM-DAT. (2016). Int. Disasters Database. http://www.emdat.be/

Femia, F. and Werrell, C. E. (2012). Syria: Climate change, drought and social unrest. *Center for Climate and Security*. https://climateandsecurity.files.wordpress.com/2012/04/syria-climate-change-drought-and-social-unrest_briefer-11.pdf

Gleick, P. H. (2014). Water, drought, climate change, and conflict in Syria. *Weather Climate and Society*, 6, 331–340. doi:10.1175/WCAS-D-13–00059.1

Hijioka, Y., Lin, E. and Pereira, J. J. (2014). Climate Change 2014: Impacts, Adaptation, and Vulnerability: Asia. IPCC 5th Assessment Report.

Hoerling, M., Eischeid, J., Perlwitz, J., Quan, X., Zhang, T. and Pegion, P. (2012). On the increased frequency of mediterranean drought. *Journal of Climate*, 25, 2146–2161. doi:10.1175/JCLI-D-11–00296.1

IRIN. (2009). Syria: Drought Driving Farmers to the Cities. IRINnews.

IRIN. (2010a). Syria: Why the Water Shortages? IRINnews.

IRIN. (2010b). Act Now to Stop Desertification, Says FAO. IRINnews.

Kelley, C. P., Mohtadi, S., Cane, M. A., Seager, R. and Kushnir, Y. (2015). Climate change in the Fertile Crescent and implications of the recent Syrian drought. *Proceedings of the National Academies of Science.*, 112, 3241–3246. doi:10.1073/pnas.1421533112

Mann, J. (2012). *The Obamians: The Struggle Inside the White House to Redefine American Power*. New York, Viking Press.

Margolis, J. E. (2012). Estimating State Instability -Extracts-Mar12–20Apr12.pdf. *Studies in Intelligence*, 56, 13–24.

Matlock, M. (2008). Water profile of Syria. *Encyclopedia of Earth.* http://www.eoearth.org/view/article/156998.

Mohtadi, S. (2012). Climate change and the Syrian uprising. *Bulletin of Atomic Scientists.* http://thebulletin.org/climate-change-and-syrian-uprising.

Nabhan, G. (2010). Drought Drives Middle Eastern Pepper Farmers Out of Business, Threatens Prized Heirloom Chiles. *Grist*, http://grist.org/article/2010-01-15-drought-drives-middle-eastern-peppers/

Paralleli. (2011). Gli Approfondimenti di Paralleli: Osservatorio Mediterraneo. *Paralleli* 3. http://www.paralleli.org/allegati/approfondimenti/osservatorio_mediterraneo/Mediterranean%20Observatory_n3.pdf

PBS. (2011). Interactive Map: Syria's Uprising. http://www.pbs.org/wgbh/pages/frontline/foreign-affairs-defense/syria-undercover/interactive-map-syrias-uprising/

Shadid, A. (2011). Syrian unrest stirs new fear of deeper sectarian divide. *New York Times.* http://www.nytimes.com/2011/06/14/world/middleeast/14syria.html?_r=0

Sternberg, T. (2012). Chinese drought, bread and the Arab Spring. *Applied Geography*, 34, 519–524. doi:10.1016/j.apgeog.2012.02.004

Sternberg, T. (2013). Chinese drought, Wheat, and the Egyptian Uprising: How a Localized Hazard Became Globalized, the Arab Spring and Climate Change. Center for Climate and Security, Stimson Center, Center for American Progress, 7-14. https://climateandsecurity.files.wordpress.com/2012/04/climatechangearabspring-ccs-cap-stimson.pdf

Sticklor, R. (2010). Syria: Beyond the Euphrates. *New Security Beat.* https://www.newsecuritybeat.org/2010/09/syria-at-the-crossroads-beyond-the-euphrates

UNHCR. (2010). *Iraqi Refugees in Syria Reluctant to Return to Home Permanently: Survey.* UNHCR. http://www.unhcr.org/en-us/news/latest/2010/10/4caf376c6/iraqi-refugees-syria-reluctant-return-home-permanently-survey.html

UNICEF. (2015). Severe Water Shortages Compound the Misery of Millions in War-Torn Syria – Says UNICEF. http://www.unicef.org/media/media_82980.html

USDA. (2008). Syria: Wheat Production in 2008/09 Declines Owing to Season-Long Drought, USDA Foreign Agricultural Service Commodity Intelligence Report. http://www.pecad.fas.usda.gov/highlights/2008/05/Syria_may2008.htm

Wadid, E., Abbashar, A. and Swaireh, L. A. (2011). Drought Vulnerability in the Arab Region: Case Study – Drought in Syria. Ten Years of Scarce Water (2000–2010). Arab Center for the Studies of Arid Zones and Dry Lands, ISDR.

Werrell, C. E., Femia, F. and Sternberg, T. (2015). Did we see it coming?: State fragility, climate vulnerability, and the uprisings in Syria and Egypt. *SAIS Review of International Affairs*, 35, 29–46. doi:10.1353/sais.2015.0002

Worth, R. F. (2010). Drought withers lush farmlands in Syria. *New York Times.* http://www.nytimes.com/2010/10/14/world/middleeast/14syria.html.

2 Post-disaster reconstruction strategies

A case study in Taiwan

Yung-Fang Chen

Increasing large-scale natural disasters, endemic vulnerability and climate change expose East Asian coastal and island nations to great environmental and social risk (UNFCCC 2007, p. 20). Taiwan, sandwiched between the Pacific Ocean and the Taiwan Straits, has recently experienced several major hazards, including the Ji-Ji Earthquake (1999), Typhoon Morokot (2009) and deadly landslides in 2010 (Hays 2013). These events highlight challenges regional governments face in preparation and mitigation of natural disasters. Significant efforts are needed to ensure inhabitants' safety, to provide effective response and to ensure the resettlement of displaced people. Governments, external agencies and relief organisations work to address the immediate needs of the affected populations. After a large-scale disaster, government decision makers are usually under great pressure to initiate reconstruction plans and projects in order to accommodate the needs of the affected populations, all while dealing with criticisms and pressure of progress made from the media (Zhang and Peacock 2009).

Emergency shelter and resettlement are a crucial part of the post-disaster phase (Lyons and Shilderman 2010; Sphere Project 2011). In particular, housing programmes, from immediate accommodation and temporary shelter through short-term residence and permanent housing, are essential to recovery (Quarantelli 1995). The increasing number of large-scale disasters in recent years has enabled academics and practitioners to share best practice (Perry and Lindell 1997; Lyons and Shilderman 2010; Sphere Project Handbook 2011). The process has led to several different challenges being identified, which include land development and logistics management. Additional issues, such as cultural conflicts, social problems and the capacity of the communities to cope, have also tested the capabilities of decision makers (Geipel 1991; Wu and Lindell 2004; Johnson 2007; Chang et al. 2011).

After the Ji-Ji Earthquake in 1999, The Taiwanese government worked to establish a more holistic disaster management system (Maa 2011); however the Morakot 88 Flood Disaster in 2009 then challenged the capabilities of the emergency response systems in post-disaster response, recovery and reconstruction (Lu et al. 2011). Although the government has invested heavily and made a serious effort to improve post-disaster housing projects, the debates on the development plans have generated tension between the government and the aboriginal groups (Quan 2010).

Both good practice and lessons learnt can be identified from these cases, and it is important that they should be shared with a wide audience both within Taiwan and across the at-risk region. In particular, the way the government works together with NGOs and the adoption of a more holistic post-disaster reconstruction framework are important aspects of this work. This chapter evaluates the post-disaster reconstruction strategies implemented after the Typhoon Morakot flood disaster in 2009. The objectives include establishing the Taiwanese context and examining how climate change has already affected the island, exploring post-disaster management strategies addressing these phenomena and providing an extensive examination of the housing projects in relation to the affected populations after Typhoon Morakot. The outcome identifies best practices relevant to other vulnerable countries in the region and makes recommendations for improved reconstruction planning and implementation.

Background of Taiwan

Taiwan is an island surrounded by the Pacific Ocean to the east, and the Taiwan Strait to the west; the East China Sea lies to the north, the Luzon Straight to the south and the South China Sea to the southwest (Figure 2.1). It is comprised of the island of Taiwan and some smaller islands, including Kinmen, Matsu and Penghu. The main island is 394 km long and 144 km across at its widest point; it has a total area of 35,873 square kilometres with a population of 23.39 million people (Ministry of Foreign Affairs 2015). Mountainous areas run north to south with a flat to rolling coastal plain running the length of the island where the majority of the population live and work. The capital city is Taipei.

A global risk analysis indicated that Taiwan is one of the most vulnerable places to natural hazards on the planet (Dilley et al. 2005, p. 8). The report summarised that 92.5 per cent of the total land and 95.5 per cent of its population are situated in areas subject to high mortality risk from multiple hazards, including typhoons, floods, mass movement (landslides) and earthquakes (Dilley et al. 2005, p. 8).

Over the years these hazards have brought about many fatalities and loss of property that have resulted in the need to resettle large numbers of affected populations. According to statistical data collected by EM-DAT (2015), from 1900 to 2015 almost 4 million people were affected, resulting in claims for more than US$213 billion. In particular, climate-related disasters (storm, flood and landslides) have affected 87 per cent of the population and account for 30 per cent of total damage reported (EM-DAT 2015).

The weather in Taiwan is mild throughout the year. On average, the temperature is 28–29°C in summer and 16–20°C in winter (Central Weather Bureau 2015). However, surrounded by the northwestern Pacific Ocean, the climate of Taiwan is strongly affected by the East Asian monsoon system. The seasonal change of wind direction also profoundly affects the weather of Taiwan. April and May are 'the Plum Rainy Season' or the so-called monsoon season. July, August and September make up the typhoon season. Heavy storms transform the short and precipitous rivers and streams into torrents of water, sweeping down the mountains carrying a large amount of mud and silt.

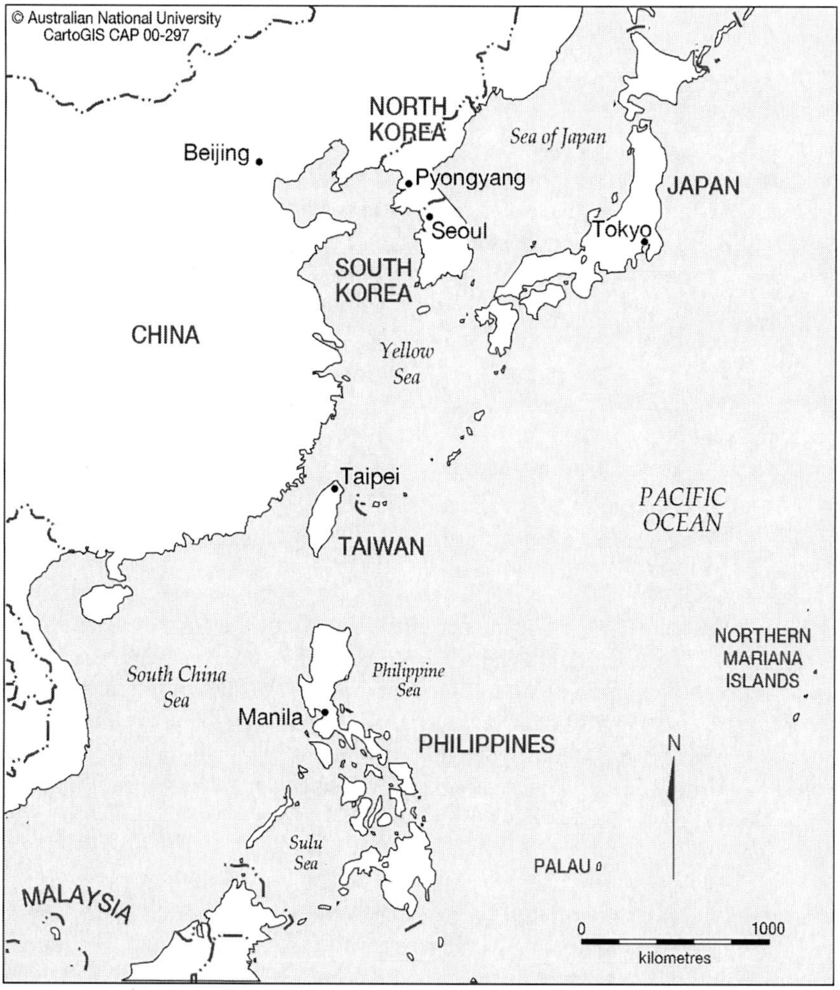

Figure 2.1 Map of Taiwan and its region

Source: CartoGIS, College of Asia and the Pacific, The Australian National University 2015

On average, three to four typhoons hit Taiwan per year. Intensive rainfall on top of the steep mountain slopes tends to trigger flash floods and landslides during the monsoon and typhoon seasons (Central Weather Bureau 2007a). Accepting the evidence of global warming research (IPCC 2013), it is indicated that there is an ongoing trend of warming in Taiwan. The National Science Council's research indicates that the average rain days in Taiwan have decreased significantly and the annual number of extremely hot days has increased (Hsu et al. 2011). It is predicted that larger and more frequent meteorological events will occur in the 21st century: for example, more heatwaves, warmer weather and heavier precipitation

(IPCC 2013, SPM-25). Although it is predicted that the mean winter precipitation on the whole island will decrease, the mean summer precipitation will increase to between 15 and 45 per cent (Wu et al. 2013). In particular, the increased number of extreme precipitation events accompanying typhoons has already had a more severe impact in the past 10 years (Wu et al. 2013).

In addition to the more vulnerable climate and environmental changes, diverse social, cultural and political factors have increased the vulnerability of the population when preparing for and responding to these potential risks. Of the 23.39 million populations, 84 per cent of them are native Taiwanese, including the Hokkien and Hakka (Ministry of Interior 2014). Mainland Chinese, Mongols, Tibetan and Manchu people, who moved to Taiwan together with the Chinese Nationalist Party, or so-called Kuomintang (KMT) from mainland China after the 1940s, became the so-called Mainlanders (Lee 2010), making up 14 per cent of the total population. Although Mainlanders are numerically the minority, they have nevertheless dominated politics from the 1940s to the 1990s. The KMT used unequal discrimination policies against the Taiwanese to continue its dominance as the ruling party (Lee 2010). Although intermarriage between Mainlanders and Taiwanese and a series of political reformations since the 1990s has gradually reduced social differences, distinct and divergent political views between the Taiwanese and Mainlanders can still be seen in the election campaigns of the two major political parties: the Kuomintang (KMT) and the Democratic Progressive Party (DPP). The inter-party rivalry and competition between these two parties has consumed many resources in the elections, which have resulted, amongst other things, in the neglect of environmental policies, urban planning and a holistic disaster management framework.

It is worth mentioning that there are also 16 aboriginal tribes who originated in the South Asian islands and who moved to Taiwan some 15,000 years ago. Although aboriginals only make up 2 per cent of the total population (CIA 2015), most of these tribes have maintained their own languages, cultures, social structures and customs whilst living in the Central Mountain Range areas. Many of these areas are classified as being at high risk of experiencing natural hazards and are designated national conservation areas. Figure 2.2 illustrates the spread of the tribes. A debate has developed as to the legitimacy of the government's risk assessment methods and development plans, since the government has restricted the activities of the aboriginals in certain, high mountainous areas. At the same time, the government has been inviting large enterprises to develop 'planned' agriculture schemes and build large-scale reservoirs. It has also been permitting logging in the forests in those same areas (Xiao 2012). This has led to controversies over the government's land use, settlement development and environmental policies, as well as its emergency management plans.

In recent years, the number of immigrants to Taiwan has been steadily increasing. In 2013, there were 586,646 residents who were not registered in a household, of whom 89 per cent were 'foreigners', 8 per cent were from the mainland, 1 per cent from Hong Kong and Macao and 2 per cent had no records (Ministry of the Interior 2014). 'Foreigners', according to the Ministry of the Interior, refers to

those who were not born in Taiwan, and who are not from Hong Kong, Macao or China. Emergency plans have paid little attention to these non-household registered residents, which makes them vulnerable if disaster strikes.

Emergency management systems in Taiwan

The Taiwanese government has been frequently challenged by tasks and responsibilities in the disaster management cycle: risk assessment, mitigation, preparation, response and recovery (Deng 2009). If, before a disaster hits, the preparation and mitigation measures are insufficiently updated and amended in time to meet the forecast changes to the climate system, then it is essential that every effort should be taken into consideration and planning for post-disaster response, recovery and reconstruction (Lu et al. 2011).

The first disaster-related law was enacted in 1965: *The Measure of Natural Hazards Mitigation in Taiwan Province*. However, the new modernised law – *the Disaster Preparedness and Response Act* (2000) – was only agreed upon due to public demand after the Ji-Ji Earthquake in the year of 1999. The Ji-Ji earthquake was the strongest earthquake in Taiwan in the late 20th century. The Act includes eight chapters and 52 articles, covering the whole range of the disaster management cycle, from mitigation and response to recovery and reconstruction. The emergency management framework can be divided into three levels: national, county and township levels.

What is notable is that the Act strengthened a more centralised approach to implementing disaster management. At the national level, the Administration Yuan, the top administration institute of the political system, takes the political responsibility. The assigned lead department has operational responsibility for different hazards. For example, the Department of the Interior is responsible for typhoons, the Ministry of Economics for flooding, and the Council for Agriculture, for landslides. The relevant leading department is responsible for calling meetings at the Emergency Operation Centre (EOC) if it foresees any emergencies.

It seems reasonable to have a lead department to coordinate a potential disaster; however, the complexity of the environment in Taiwan means that if there is a typhoon, flood and landslides will ensue. Quite often this results in undefined 'ownership' of the disaster and the arrangement becomes inefficient when dealing with complexities (Professional Advisory Committee of Disaster Reduction 2012a, pp. 7–8). What is worrying is that if a risk is not predefined, none of the departments will take the lead, due to an organisational culture of 'not getting involved to avoid trouble' (Maa 2011). Consequently, there has been a call for an 'all-hazard approach' to manage the disaster management structure. In addition, the complexity of the organisational management structure at the national level led to paralysis of the decision-making and the command and control mechanism during a disaster event. Scholars have been calling for the creation of an all-hazards unified disaster management (Tsai and Chi 2010; Maa 2011).

The centralisation of the emergency management structure challenges the capability and capacity of the local authorities. Centralisation means that the national government is in charge of the budget plan and of coordinating the tasks of the local governments to deliver and implement the arrangements for disaster management. The wide range of tasks assigned to the local government meant that it was not possible to complete all of the missions (Sa 2010). At the same time, the centralised structure slowed down the post-disaster recovery and reconstruction process, due to the lengthy administrative procedures. After Typhoon Morakot in 2009, an attempt was made to introduce the district office as the fourth level of authority to implement disaster intervention and response strategies, in order to enforce the responsibilities at local level.

The reconstruction section in the *Disaster Preparedness and Response Act* (2000) requires the government to include social, economic, resettlement, infrastructure and other development dimensions in the reconstruction programmes. The government at all levels should be responsible for recovery and reconstruction after a disaster has struck. The Act also encourages utility companies and civic societies to assist with the implementation of the strategies. However, after the blast disaster in a Water Park in New Taipei City in June 2015, there have been calls for a more detailed explanation of the recovery and reconstruction, in particular, regarding the compensation and relief fund.

Impact of Typhoon Morakot

Typhoon Morakot (2009) was one of the most destructive typhoons ever to impact on Southern Taiwan (Tsou et al. 2011). The typhoon formed on August 5, 2009, and made ground fall near Hualien City at 23:50 on August 7, before leaving Taiwan near Taoyuan at 14.00 on August 8 (Chen 2009). Although the scale of the typhoon was classified as a medium-strength event according to the grading system of the Taiwan Central Weather Bureau, its duration within the island triggered many serious floods, landslides and debris flows in the remote and thinly populated mountainous areas (Chen 2009). Major routes to many remote villages were cut off; in total, 10 isolated areas in the mountainous regions and one heavily flooded area in the coastal regions were affected (Chern and Hung 2012). Interrupted and broken roads and bridges prevented search and rescue teams from reaching many of these areas. Many households lost electricity and telecommunication systems.

The Morakot Post-Disaster Reconstruction Council (2014) estimated that the typhoon accounted for almost 80 per cent of the annual expected rainfall, which caused major flooding and accompanying landslides. It resulted in 699 deaths; nearly 40 per cent of the homes of the population were either destroyed or damaged. The estimated financial loss from the typhoon was nearly US $61.17 billion; for details see Table 2.1 (Morakot Post-Disaster Reconstruction Council 2014).

Table 2.1 Estimate of losses from Typhoon Morakot

Item	*Amount (US$ million)*
1. Total property damage that can be monetized	64.44
(1) Direct loss (loss of property)	61.17
a. Flood damage to homes	1.71
b. General damage to homes and buildings (including amenities)	1.4
c. Direct industry losses	8.82
(a) Agriculture, forestry, fishing and livestock industries	6.26
(b) Industrial (direct losses to business)	0.75
(c) Commercial (direct losses to specialty business districts)	0.38
(d) Tourism facilities	0.7
(e) Indigenous specialty industries	0.73
d. Infrastructure damage	49.23
(2) Indirect losses (loss of revenues)	3.27
a. Agriculture	2.63
b. Industry	0.19
c. Service industry	0.45
2. "Losses that cannot be monetized: 699 deaths or missing, damage to natural environment and scenic areas, damage to indigenous people's culture, humanities and historic sites."	

Source: Morakot Post-Disaster Reconstruction Council 2014, p. 18

The response phase started very early, well before Typhoon Morakot landed on the island. The Emergency Operation Centre (EOC) at the national level was activated on the evening of August 5, 2009 after the Weather Bureau issued the sea typhoon alert. When the severe weather alert was issued at 8.30 on August 6, key responders were on call to support decisions made at the EOC. As the forecast of the route of the typhoon and heavy rainfall focused on northern Taiwan, all attention was paid to this part of the island.

However, the unexpected heavy rainfall in Southern Taiwan challenged the capability of the emergency response systems, in particular, the coordination and communication mechanisms within these emergency management plans. Incorrect forecasting and the breakdown of the mobile/telecommunication network prevented the decision makers from understanding the development of the disaster. The Southern Typhoon Coordination Centre was only activated on August 9 in Ping-Dong County; at the same time, a Tactical Command Centre was opened in Qi-Shang to support search rescue, logistic supply, temporary shelter and reconstruction (Morakot Post-Disaster Reconstruction Council 2012b, pp. 49–50).

Short-term and medium-term settlement post Morakot

In the first post-Morakot disaster phase, the Taiwanese government established 170 emergency shelters, including temples, churches, schools, community centres and farms to accommodate affected populations. In total, 8,189 individuals of the affected populations used these facilities (Morakot Post-Disaster Reconstruction Council 2012a, p. 44). Learning from the experience of the Ji-Ji Earthquake of 1999, the government very effectively mobilised the emergency shelters and then quickly moved to the temporary housing phase.

Relief funds were made available so that subsidies could be provided to the affected populations either for renting houses, living in government-organised pre-fabricated housing or military barracks/veterans homes. Some people chose to purchase new homes in the private housing sector. Others took out loans in order to repair their houses (in-situ reconstruction). Approximately 4,600 of the affected populations were allocated to military barracks accommodation. NGOs such as the Red Cross and World Vision started to help to build pre-fabricated houses as mid-term temporary shelters (Morakot Post-Disaster Reconstruction Council 2012a, p. 49). In addition to these arrangements, relevant recovery plans for critical infrastructure repairs, community facilities and industrial reconstruction were made. Long-term reconstruction plans were also considered. The full resettlement strategy is displayed in Figure 2.2. Research shows that Taiwan is among the countries which recover faster from disasters than many others, due to its economic advantages (Aldrich 2012).

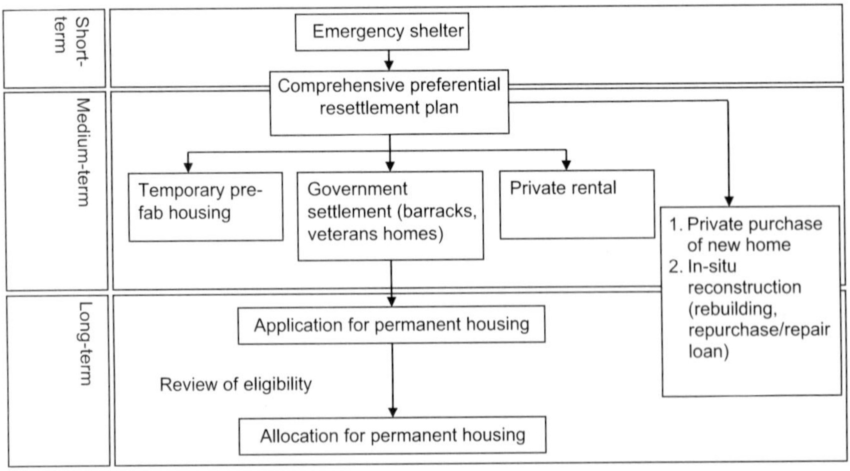

Figure 2.2 Government alternative resettlement approaches provided with the affected populations

Source: adapted from Morakot Post-Disaster Reconstruction Council 2011, p. 28

Change of reconstruction strategy – the introduction of the one-stop housing policy

The post-disaster impact assessment revealed the difficulties of reconstruction. In total, 1,766 houses were seriously damaged and more than 160 inhabited areas were unsafe to live in (Chern 2011). At the same time, 6,316 households (19,191 residents) lived in the newly defined high risk areas (Chern 2012).

According to the Disaster Preparedness and Response Act (2000), the government has the responsibility to help rebuild or reconstruct housing for the affected populations. Although the government recognised the necessity of utilising a mid-term settlement strategy to accommodate those affected, it was difficult to find sufficiently large areas of available land to accomplish both pre-fabricated and permanent housing, due to the number of those affected.

To overcome the challenge, a 'one-stop housing policy' was proposed for people whose house had been destroyed and needed to be 'relocated.' It was required that new, permanent housing should be located on the original site, wherever possible, provided it was safe, or else it should be relocated to the nearest suitable location. Several justifying factors were cited at the time. First, it would reduce the total cost of reconstruction because there would be no need to invest in pre-fabricated houses. Second, it would alleviate the difficulties of finding appropriate land for the large number of affected populations needing to be temporarily re-housed. Third, some believed that if permanent houses were rapidly built in a short space of time, the social and security issues inherent in temporary housing schemes could be avoided. Finally, the additional costs of management and demolition expenses could also be saved (Morakot Post-Disaster Reconstruction Council 2011).

The new permanent houses were built in only six months and were free of charge for those who were eligible to apply for one. Although it was believed to be a great financial support for those affected populations, the 'one-stop housing policy' raised much controversy and debate.

Unreliable post-disaster impact assessment

First of all, the reliability and validity of the post-disaster impact assessment was challenged. To speed up the process of resettlement, the government used a 'fast-track' approach to conduct environmental risk assessments when defining appropriate sites for the newly built permanent houses. The 'regulations for zoning and selection of re-settlement site and construction' (1999) indicated that relevant authorities could conduct the site investigation without consulting other relevant agencies or local residents. As a result, many investigations were based solely on a few experts' experience and on aerial images (Lu 2013, p. 258), so the recognition of the degree of damage, and the necessity for relocation, was seen as subjective and has become controversial. Regardless, after the assessment, it was concluded that 3,096 households had to be moved to permanent housing, 88.7 per cent of them had to leave their original village, and most of them were aboriginals (72.5 per cent) (Morakot Post-Disaster Reconstruction Council 2011).

Is relocation necessary?

The new zoning and relocation policy was ambiguous and confusing to the affected populations due to a lack of effective communication between the government and the public. Since the policy was decided very quickly after the disaster struck, many affected populations did not have a chance to return home to inspect the damage to their property. Some still believed that it was safe for them to return to their homeland. In addition, they had not been informed as to whether or not they could return or stay in their homes after being relocated. Nor were they told what activities they could undertake in their original homes after being relocated. Many affected populations were not prepared to leave their homes without further discussion about their needs and wishes (Quan 2010). These issues led to more controversies in the communities that had to be relocated and to major public distrust of the government (Ke 2010; Lu et al. 2011).

Although most of those who had to leave their original villages were aboriginals, most of those affected populations actually welcomed the relocation policy on the condition that the government would discuss with the affected communities their future development (Quan 2010). Those who preferred to move to the newly built community felt that it was an opportunity to obtain a better education for their children and it would also reduce the cost of running a family. One resident who had moved to a rebuilt community said,

> In the past we needed to have two houses, one in the mountains and one in the city. To provide children with better education, we needed to buy or rent a house in the city. At the same time we were working in the mountains, so, we needed to make provision for two sets of household goods: for example, gas ovens, washing machines and refrigerators. It is a very heavy burden to maintain two houses, and therefore the quality of life was decreasing. Now we only need to maintain one house and consequently we have a better life.
> (Knight et al. 2012, p. 177)

Those who preferred to stay in the original village, however, would trade the environmental risks against other reasons, such as maintaining ancestral connections, as well as economic factors. The relocation regulations did not consider or take into account the relevant or consequent impacts of the relocation. These include the amount of assistance the affected population would receive from the government, the types of house they would live in after they were relocated and many other practical consequences of the move. The affected population felt that it was a way the government could force them to leave the mountains where they had lived as farmers tending their crops for generations (Ke 2010). Shieh et al. (2013) argued that, although a free permanent house was to be given, without any farmland or the possibility of returning to their original homelands, it would be the same as extinguishing the tribe.

Since many of the affected population felt that the government did not pay attention to the welfare of the aboriginals, strikes were organised to make sure the

reconstruction strategies supported the development of aboriginal communities. After a standoff lasting almost a year against those who did not wish to move, the government finally agreed that aboriginals could stay in and/or return to their original houses/locations, with a condition that a nearby emergency shelter be prepared.

The reconstruction of the original communities

Normally the reason given for relocation is that the houses, due to the level of damage, cannot be rebuilt; in this case, most of the houses in the newly determined high risk areas had not been destroyed and could be repaired. They were classified as high risk areas only because the surrounding environment had become dangerous and/or the main traffic routes to the community were destroyed, resulting in difficulties in communicating with the external environment (Lu 2011). It is not easy to justify the decision of 'compulsory resettlement' and 'village relocation' (Shieh et al. 2013), particularly within the context of the long history of debates between the government and the aboriginals on the use of mountainous land.

The government preferred not to fund those areas susceptible to repetitive disaster on grounds of safety and environmental protection. In addition, it could be cheaper to provide free housing than to rebuild the key infrastructure. However, residents who preferred to stay in their own homeland felt that it is a basic human right to ask the government to provide them with the essential facilities, so they could return to and continue to live in their homelands. Finally, the government started to provide limited support and the construction of major roads in these high risk areas.

Difficult to recognise the eligible applications

Another challenge for the relocation policy was the eligibility criteria. Each affected household could apply for a permanent house if their original home was recognised as being collapsed, having an approved relocation household or the original house being located in the newly determined settlement areas (Chern and Liu 2011, p. 68). To be eligible to apply for a permanent house, the affected population had to provide evidence of their occupation of the original house. However, as a result of the disaster, many of the residents had lost the necessary evidence or the deeds to their homes or lands. Since most of the affected populations were aboriginals, their land tenure rights had become highly complicated, with some deeds stretching back to the Japanese colonial period. These major complications had prevented them from providing the necessary proof and had thus postponed the government procedures for applying for the permanent houses (Quan 2010).

The government decision makers finally agreed to adopt a 'collective relocation' approach to overcome the issue. It was thought that if communities could reach an agreement to be relocated, the residents could then gain permission to have a permanent house, and it would not be necessary to provide individual evidence. Unsurprisingly, it was not easy to reach universal agreements amongst

entire communities; hence, this decision created further disagreement among those affected.

The other challenge when an application was successful was to determine the precise number of occupants within each household. It was decided that the affected populations could apply for different types of houses: 46.2m^2 for a family of two, 92.4m^2 for three to five people and 112.2m^2 for 6–10 people, according to the records shown in the 'Household Registration Book' before the disaster. Many aboriginals sent their children to live in urban areas to receive a better education, hence the records of their children were registered with their resident house in the urban area. It therefore became a serious problem to recognise the 'true' number of people within each household. The residents had to negotiate with NGOs to determine the size of the houses, if the NGO was willing to risk ignoring the government's strategy. Since the Taiwan World Vision organisation had been working in the aboriginal communities for decades, they were relatively more open to this kind of discussion.

Working with NGOs to construct new communities

It was the first time that the Taiwanese Government had liaised with local authorities and NGOs to implement the construction work necessary after a disaster. Once the relocation sites were decided, the government worked with NGOs such as the Tzu-Chi Foundation, Taiwan World Vision, Taiwan Red Cross, Dharma Drum Mountain and others to build permanent houses. The NGOs were responsible for building the houses while the government was responsible for the establishment of key infrastructure in these newly built communities (Table 2.2).

The government and NGOs had different agendas with different priorities. The former wanted rapid results so NGOs were required to spend the funds allocated to reconstruction projects within three to four years. They had to apply for funding every year. However, reconstruction work could take up to 10 years or even longer. NGOs had therefore to speed up the reconstruction process, despite their organisation's policy on the reconstruction. In addition, NGOs had to find their own sources of funding if they wished to provide more consistent support to the communities. As a consequence, it was inevitable that NGOs had, at times, a strained relationship with government departments and institutes.

Table 2.2 The responsibility/deliverables for the respected lead agencies

Lead agency	*Responsibilities*
Central Government	Policy, law, budget, land offer, reduce procedures matching resources, major infrastructure construction
Local Government	Qualification review, building administration, community management
NGOs	Fund raising, buildings and local infrastructure construction

Source: Chern 2012, p. 54

NGOs also had a different scale of human, financial and material resources. For example, compared to other NGOs, Tzu-Chi had obtained substantially more donations and had also recruited more volunteers to help on the construction projects. This allowed the organisation to build the first new community of 628 households within 88 days (Chern 2012). In contrast, other NGOs such as World Vision and Dharma Drum Mountain had been struggling with their finances and could only support the building of a limited number of communities.

Finally, previous experience and already established networks within the affected communities influenced the effectiveness of the reconstruction work. Social trust was an important factor in the reconstruction of affected communities. As has previously been said, Taiwan World Vision has been working in the aboriginal communities for a long time; they are more familiar with the aboriginal culture, society and lifestyles. Therefore, if any disaster occurred in the aboriginal communities, the government would ask World Vision to help coordinate operations. On the other hand, if a disaster occurred in a non-aboriginal area, it would be the Red Cross or Tzu-chi who would be asked to help. It must always be borne in mind that although most of the victims were from the aboriginal population, since the scale of the disaster was so huge, it was not possible for only one NGO to be responsible for all reconstruction work.

The use of a people-centred approach

The Taiwanese Government and the responsible NGOs made a laudable attempt to use a people-centred approach to deliver the reconstruction projects, by, for example, holding consultation meetings with the affected population to discuss the methods for reconstruction. Affected populations were also employed for the construction projects. However, several issues and difficulties were identified.

First, the decisions on the reconstruction policy, in particular, the Morakot Typhoon Post Disaster Reconstruction Act, were made very quickly so no public hearings or consultations were held. Publicly elected mayors of the affected towns and villages participated in the policy-making procedures, although affected populations did not necessarily find them representing the opinions of their communities. In addition, the qualification and suitability of the members in the meetings in the Legislative Yuan were debatable (Huang 2012).

Second, although NGOs tried to consult the affected population during the construction process, if a NGO had to meet the project deadlines proposed by the government, it had to speed up the consultancy process with the affected population. This meant that some of the participants might not have understood or been clear about the messages that were delivered in the consultation meetings, and some public concerns would either not be discussed, or would be ignored.

Since NGOs took the lead in the construction process, they recruited professional interior designers and urban planners to plan for the houses and the community. However, this was more expensive, and even then the quality of the houses and their interior design might not be suitable for residents' needs. The factors described above resulted in a low occupation rate (Lu 2013).

Lu et al. (2013) also claimed that the strategy caused more disputes in the long-term reconstruction as the residents did not have many opportunities to discuss and participate in the reconstruction projects (Lu et al. 2013).

All-inclusive community development – the Rainbow Communities Programme

Though there were many controversies in the relocation projects, the Taiwanese government attempted to facilitate the use of an all-inclusive approach for community development to ensure the sustainability of the livelihood of the relocated populations, namely the Rainbow Communities Programme. Once the new houses were built, the government continued to work with NGOs to establish 'Life Reconstruction Service Centres' to help the affected populations rebuild their ways of living as well as their livelihoods. The government provided grants for NGOs to provide services to the community residents (Feng 2011). The services provided by these Centres included mental health, education, employment, welfare, community development and citizen advice facilities (Table 2.3).

The government also set up priorities for the work required: starting with the needs assessment of each individual household, then to facilitate the development of each individual's priorities and requirements and finally to promote the development of the whole community. Training and development plans were provided for these communities.

The government claimed success in the implementation of these projects (Morakot Post-Disaster Reconstruction Council 2012b), for example, the satisfaction survey conducted by the Ministry of Interior in 2010, showed that 89 per cent of the residents were satisfied with the permanent housing, and 85 per cent were happy with the overall quality of life. Nevertheless, there was a continually declining trend of satisfaction. The results from Knight et al. (2013) showed that only 76 per cent of the population was satisfied with the condition of their houses, and 30 per cent of the residents were not happy with the way the government had

Table 2.3 Six dimensions of service provided by the Life Reconstruction Service Centre

Mental	Mental consultancy and medical transfer assistance
Education	Student schooling and counselling assistance
Employment	Assisting the jobless to apply for unemployment payment, attend professional training and apply for jobs
Welfare	Sustaining needs of elders, youngsters, single-parent or low-income families
Community development	Helping create local job opportunities and enhance the development of local industries
Citizens' advice facilities	Transferring legal, community industry reconstruction, cultural display news

Source: Chiayi County Government 2015

worked with the NGOs in the promotion of the reconstruction and development programmes. It was noted that the programmes were based on the government's annual funding so that the services provided by the NGOs tended to be of a temporary nature. NGOs had to write a proposal each year to make sure the projects could be continued.

In addition, some aboriginals did not feel that the government had provided them with the right support, and this left them struggling with finding ways to adapt to a new life and to settle into new types of occupations. For example, with the government-funded courses to train the residents to utilise computers and learn marketing skills, residents could not relate these skills to their livelihood and career development. As a consequence, many of those residents have gradually returned to their original houses in the mountains and used the newly built houses as 'emergency shelters', From a positive point of view, it is argued that at least these residents keep a safe place in an urban area in case of typhoon, so they will be more willing to be evacuated (Quan 2010). These development programmes have lasted five years and most are now complete, although not all communities are back to 'where they were,' and indeed, some are worse off.

Recommendations

The chapter has reviewed the arrangements that the Taiwanese government made to assist the population affected by Typhoon Morakot. Debates, controversies and arguments have all been illuminated in the previous sections. This section attempts to summarise several recommendations to bear in mind for future programmes of reconstruction, consequent to any further disasters.

1 Plans for reconstruction should start to be considered during the immediate disaster response phase. Worst-case scenario should be included in the emergency management plans.

Although many reports identifying lessons learnt have been published after the disaster, the main recommendations remain focused on the pre-disaster mitigation and preparedness and the revision of the disaster management framework. For example, Deng (2009) said that little consideration has been put in the decision-making procedures. The Professional Advisory Committee of Disaster Reduction (2010, pp. 47–48) identified three points to take into account for the reconstruction programme and Lu et al. (2013) focused mainly on the social vulnerability aspects.

However, the lack of a formal settlement and reconstruction plan to cover for catastrophic events prevents effective management at the beginning of a climate disaster. The worst-case scenario should at all times be taken into consideration to ensure a holistic plan can be formed. It is recommended that the recovery and reconstruction plans should be included during discussions on handling the recovery phase.

2 *Using insurance to transfer risks.*

It is possible, perhaps inevitable, that, due to climate change, more frequent and even larger scale future disasters will be encountered in Taiwan. This means that the government should consider other alternatives for risk transfer. Since climate change will introduce more drought and more flood in Taiwan, it is necessary to consider more holistic ways to introduce insurance policies to the public and invite insurance companies to become more involved in the framework.

Currently it is not mandatory to purchase natural hazard insurance. Some people only have earthquake insurance, but few have flood and other insurance protecting them from the consequences of natural hazards. It is recommended that there should be an all-inclusive insurance policy introduced and promoted to cover potential earthquake, flood, landslide, typhoon and house fire damage (Professional Advisory Committee of Disaster Reduction 2010). It would be a responsibility of the government to promote such policies. With lower income families, it is recommended that some sort of micro-insurance policy be introduced which would ensure a minimum level of protection.

3 *Ensure valid public participation.*

Many criticisms have been raised by the affected aboriginal population. Their argument was that they had different cultures, beliefs and lifestyles so decision makers should work with their needs and demands in mind when informing them of any new policies (Quan 2010; Huang 2012). In this case, the main issue was the consultation procedure during the construction phase. The government should provide channels for discussion at a very early stage, thus allowing time for in-depth and considered discussion. At the same time, a programme of educating the public on the concept of 'public participation' should be implemented.

Another problem was that the public were not aware of any new potential risks after Morakot struck. The foundation of risk communication lies in whether information delivered has been received and been acted upon. The communication flow should not only descend from top to bottom, but also rise from the bottom up. Risk communication and reliable information between the government and the public during normal time is essential in order to build the required level of trust and confidence between government and public.

4 *The impact of relocating communities into the existing communities.*

More research is needed into the effect on the resettled communities and their surrounding areas. First, the long-term impact on the relocated residents should be examined. Although the government has created several training programmes to learn new skills, it is necessary to observe the long-term consequences on people who have undergone a traumatic experience, for example from psychological, social and economic perspectives.

Second, the long-term impact on nearby existing communities should also be monitored. Research has rarely examined how relocated residents influence and have an effect on those residents who live near a newly built community. A study of the impact of any big reconstruction or new construction project on the formal and informal social networks of the surrounding area is required.

It is unavoidable that a newly settled population would have some kind of interaction with the existing residents, for example, shopping, parking and leisure activities. The new residents carry with them their original cultures, beliefs and behaviours, and this will inevitably impact the existing residents. They will also create a new social network amongst themselves, so it may not always be easy for them to integrate with their new neighbours. It is necessary to study these changes and/or the formation of new social connections, taking into account also the economic and security impacts on the surrounding neighbourhoods.

Discussion/conclusion

As climate-related disasters increase in Asia (EM-DAT 2012), the impact of these events has become greater (UNISDR 2015, p. 50). The recent Morakot disaster 2009 in Taiwan is a typical example of such disasters that occur in many parts of the world. The lessons identified from the post-Morakot response, recovery and reconstruction are directly applicable to other countries and regions. Whilst disaster mortality has decreased, response, recovery and reconstruction remain major political and socio-economic challenges.

Experience gained from several large-scale housing reconstruction programmes has been vital to improving disaster response. For instance, housing recovery after the 2006 Yogyakarta Earthquake in Indonesia, the Colombia Earthquake in the coffee-growing region (Armenia) and the Pakistan rural housing reconstruction programme after the 2005 earthquake identified that the reconstruction would take more than two years to complete. The World Bank has identified community-driven and owner-driven reconstruction as the best practice for housing reconstruction (World Bank Group 2014, pp. 16–18). However, the issues of land use, funding and the psychological health of the affected population have challenged many reconstruction projects such as the reconstruction projects in Tohoku, Japan (Fuji 2015). After the impact of Typhoon Morakot in 2009, the Government of Taiwan utilised an effective approach to the reconstruction and rehabilitation of the affected properties and populations. Working with NGOs and the introduction of the one-stop housing policy enabled a large proportion of the affected population to have permanent houses a very short time after the disaster struck. At the same time, the use of all-inclusive community development, exemplified by the Rainbow Communities programme, has been successful. New lessons can be identified and shared with a wider audience.

Planning for post-disaster recovery and reconstruction before an event takes place can effectively enhance community resilience (Johnson 2014). Recommendations in this chapter identify strategies for recovery and reconstruction that should be embedded in the post-event response phase. Since the characteristics and

impact of each disaster is different, decision makers should consider the worst-case scenario for each type of disaster and make strategies to intervene appropriately to minimise the impact. The UK government's publication 'Emergency Response and Recovery' stresses the importance of starting the recovery phase alongside the immediate response in order to meet the demands of those affected (Cabinet Office 2013, p. 83).

Financial arrangements, including compulsory hazards insurance policies, should be considered for property owners as these are an effective risk transfer tool and could further facilitate reduction measures (Twigg 2015). It might be difficult to apply a national insurance plan, such as the US national flood insurance scheme, to a high risk area due to the unwillingness of insurers to cover repetitive loss and compensation claims (King 2005; Kron 2008), although it is possible to impose micro-insurance policies to transfer risks for low-income households and businesses. Successful experiences include India, Nepal, Bangladesh, Pakistan and Malawi (Mechler et al. 2006).

Participation and engagement with the public can be a powerful instrument when the affected populations express their needs and requirements prior to implementing reconstruction projects (UNDRO 1982). Channels for discussion should be opened at a very early stage of the reconstruction process, thus allowing time for in-depth and considered discussion. Although Taiwan has been ranked as the third most democratic country in Asia in 2014 (Economist Intelligence Unit 2015), the experience in post-Morakot reconstruction nevertheless suggests that a more effective risk communication channel should be built. While there are different levels of community participation (Arnstein 1969; Davidson et al. 2007), it is suggested that widespread programmes of educating the public on the concept of 'public participation' should be implemented as widely and as soon as possible. Such programmes could be useful not only for reconstruction but to enable disaster risk reduction measures to be introduced.

Taiwan is located in a very high risk area, but the public and the government has long been implementing a wide range of measures to reduce potential risks and the impact of any disasters. Although it is possible to improve operations in the mitigation and response phases, comparably, the recovery and reconstruction phases seem to be more considerate. This chapter first illustrates the impacts of climate change on the island of Taiwan, and then it focuses on the discussion of the housing strategies utilised after Typhoon Morakot. Best practices and recommendations are made and are shared with countries in other regions in an attempt to share the experience so as to improve the reconstruction plans. Despite the improvements made, there is still a need to advance systems to include reconstruction planning, public participation, an addressing of the impacts of relocating communities, insurance for transferring risks and preparation for worst-case scenarios.

However, the speed and effectiveness of recovery and the holistic reconstruction approaches should be shared, in particular between the effective coordination and collaboration between the government and the NGOs. A clearer definition of the role and responsibilities between the government and NGOs would enable

more effective collaboration between these two sectors. In addition, the Taiwanese government proactively used an inclusive approach to facilitate reconstruction and development, including the growth of infrastructure, community resilience and individual skills and knowledge. What is noted is that, while most reconstruction studies focus on the rebuilt communities themselves, there is also a need to study the impact on the surrounding neighbourhoods of major works or new construction projects. There can be profound effects to existing, nearby communities on both the formal and informal social networks, taking into account the effect on the economic, social and security impact.

References

Aldrich, D.P. (2012). *Building Resilience: Social Capital in Post-Disaster Recovery*. The University of Chicago Press, Chicago and London.

Arnstein, S.R. (1969). A ladder of citizen participation. *AIP Journal*, 35(4), 216–224. https://www.planning.org/pas/memo/2007/mar/pdf/JAPA35No4.pdf.

Cabinet Office. (2013). *Emergency Response and Recovery: Non Statutory Guidance Accompanying the Civil Contingencies Act 2004*. Cabinet Office, London. https://www.gov.uk/government/uploads/system/uploads/attachment_data/file/ 253488/Emergency_Response_and_Recovery_5th_edition_October_2013.pdf.

CartoGIS, College of Asia and the Pacific, The Australian National University. (2015). Map of Taiwan Region. http://asiapacific.anu.edu.au/mapsonline/base-maps/taiwan-region.

Central Weather Bureau. (2007a). Natural Hazards Q&A. http://www.cwb.gov.tw/V7/prevent/plan/prevent-faq/ (Accessed: 6 November 2014).

Central Weather Bureau. (2007b). Earthquake Hazards. http://www.cwb.gov.tw/V7/knowledge/planning/seismological.htm (Accessed: 6 November 2014).

Central Weather Bureau. (2015). Brief of the Climate of Taiwan – The Temperature in Taiwan. http://www.cwb.gov.tw/V7/climate/climate_info/statistics/statistics_1_3.html?.

Chang, Y., Wilkinson, S., Brunsdon, D., Seville, E. and Potangaroa, R. (2011). An integrated approach: Managing resources for post-disaster reconstruction. *Disasters*, 35(4), 739–765.

Chen, Y.L. (2009). Investigation of Typhoons in 2009 – A Report on No.8 Typhoon Morakot (0908). Central Weather Bureau. [陳怡良，民國98年颱風調查報告 – 大8號莫拉克颱風 (0908)。中央氣象局氣象預報中心。]. http://photino.cwb.gov.tw/rdcweb/lib/cd/cd02tyrp/typ/2009/98_02_Morakot_0908.pdf.

Chern, J.C. (2011). Morakot Post-Disaster Reconstruction in Taiwan. *Presentation for the international conference on post-Morakot homeland reconstruction with sustainable development: Rebuilding sustainable homeland with innovation and united efforts*. Kaohsiung, Taiwan, 30 July 2011.

Chern, J.C. (2012). An Innovative Approach for Post-Disaster Reconstruction of Typhoon Morakot in Taiwan. *Presentation for the 8th international symposium on social management systems*. Kaohsiung, Taiwan, 2–4 May 2012.

Chern, J.C. and Hung, S.Y. (2012). The strategy of community reconstruction after a mega disaster: The case of Post-Typhoon Morakot Reconstruction in Taiwan. *Journal of Management*, 1(1), 63–79.

Chern, J.C. and Liu, C.Y. (2011). Efficiency, respect and innovation in restoration. *Taiwan Architect Magazine*, 444, 68–75. [陳振川、劉敬宗(2011)莫拉克颱風災後重建——效率、尊重、創新。建築師，444 卷：頁 68–75。].

Chiayi County Government. (2015). Disaster Prevention and Rescue Management & Post-Morakot Reconstruction. *Presentation*. Chiayi Government, Chiayi County, Taiwan. 29 May 2015.

CIA. (2015). *The World Fact Book: Taiwan*. https://www.cia.gov/library/publications/the-world-factbook/geos/tw.html.

Council of Indigenous People. (2010). Spread of Aboriginal Tribes in Taiwan. http://www.taiwan.gov.tw/ct.asp?xItem=27735&ctNode=1918&mp=1001.

Davidson, C.H., Johnson, C., Lizarralde, G., Dikmen, N. and Sliwinski, A. (2007). Truths and myths about community participation in post-disaster housing projects. *Habitat International*, 31, 100–115.

Deng, Z.Z. (2009). An Investigation on the Development of the Disaster Management Framework in the Upcoming 10 Years. Project funded by National Fire Agency. [鄧子正(2009)未來十年我國災害管理發展趨勢及因應策略之研究。內政部消防署委託研究報告。].

Dilley, M., Chen, R.S., Deichmann, U., Learner-Lam, A.L., Arnold, M., Agwe, J., . . . Yetman, G. (2005). *Natural Disaster Hotspots: A Global Risk Analysis*. Disaster Risk Management Series No 5. World Bank, Washington, DC.

Drabek, T.E. (1986). *Human System Responses to Disaster: An Inventory of Sociological Findings*. Springer-Verlog, New York.

Economist Intelligence Unit. (2015). Democracy Index 2014 democracy and Its Discontents: A Report from the Economist Intelligence Unit. http://www.sudestada.com.uy/Content/Articles/421a313a-d58f-462e-9b24–2504a37f6b56/Democracy-index-2014.pdf.

EM-DAT. (2015) and EM-DAT. (2012). Natural Disasters Trend. World 1900–2011. http://www.emdat.be/natural-disasters-trends. The OFDA/CRED International Disaster Database – www.emdat.be – Université Catholique de Louvain – Brussels – Belgium.

Executive Yuan. (2014). Introduction of Republic of China: International Trade. http://www.ey.gov.tw/state/News_Content3.aspx?n=1DA8EDDD65ECB8D4&sms=474D9346A19A4989&s=8A1DCA5A3BFAD09C.

Feng, Y. (2011). Practical and representative post disaster reconstruction policies and strategies: The development of the affected populations' welfare in Taiwan. In Zhang, Z. Z. (ed.), *Social Welfare Models – From Tradition to Innovation*. Chinese Association for Relief and Ensuing Services, Taipei. 249–262. [馮燕 (2011) 具有可行性與回應性的災後重建政策與法規——災民福利法規在台灣的發展。張正中編輯，社會福利模式——從傳承到創新。台北：中華救助總會: 249–262。].

Fuji, N. (2015). *Psychological Consequences Remain Profound in Coastal Areas of Tohoku*. News, Tohoku University. http://www.tohoku.ac.jp/en/news/research/news20150901.html.

Geipel, R. (1991). *Long-Term Consequences of Disasters: The Reconstruction of Friuli, Italy in Its International Context, 1976–1988*. Springer-Verlag, New York.

Hass, J.E., Kates, R.W. and Bowden, M.J., eds. (1977). *Reconstruction Following Disaster*. The MIT Press, Cambridge, MA.

Hays, W. (2013). Lessons Learned from Past Notable Disasters, Taiwan. Part 2: Typhoons, Floods and Landslides. *Power point presentation disasters supercourse (natural and manmade)*. http://www.pitt.edu/~super1/collections/collection52.htm.

Hsu, H.H., Chou, C., Wu, Y.C., Lu, M.M., Chen, C.T. and Chen, Y.M. (2011). Climate Change in Taiwan: Scientific Report 2011 (Summary). National Science Council, Taipei, Taiwan, ROC. http://tccip.ncdr.nat.gov.tw/v2/upload/book/20150410100030.pdf.

Huang, Z.H. (2012). The vulnerability of pluralism: The philosophy, legislation and organisational structure of post Morakot reconstruction. In Pan, Y.H. (ed.), *Conference*

Proceeding: The Re-Introduction of the Culture of PingPu Tribe in Southern Taiwan. National Taiwan Museum, Taipei. [黃智慧 (2012)「多元文化」理念的脆弱性：莫拉克災後重建政策思維、法令與組織型態。潘英海主編，再現南台平埔族群文化學術研討會論文集。台北：台灣博物館。] http://idv.sinica.edu.tw/etwisdom/2009Web/PDF/20120820.pdf.

IPCC. (2013). Summary for policymakers. In Stocker, T.F., Qin, D., Plattner, G.-K., Tignor, M., Allen, S.K., Boschung, J., Nauels, A., Xia, Y., Bex, V. and Midgley, P.M. (eds.), *Climate Change 2013: The Physical Science Basis: Contribution of Working Group I to the Fifth Assessment Report of the Intergovernmental Panel on Climate Change*. Cambridge University Press, Cambridge, UK and New York, NY, USA. http://www.ipcc.ch/pdf/assessment-report/ar5/wg1/WG1AR5_SPM_FINAL.pdf (Accessed: 6 November 2014).

Jha, A.K., Barenstein, J.D., Phelps, P.M., Pittet, D. and Sena, S. (2010). *Safer Homes, Stronger Communities: A Handbook for Reconstructing after Natural Disasters*. World Bank, Washington, DC.

Johnson, C. (2007). Strategic planning for post-disaster temporary housing. *Disasters*, 31(4), 435–458.

Johnson, L.A. (2014). Chapter 6 long-term recovery planning: The process of planning. In Schwab, J. C. (ed.), *Planning for Post-Disaster Recovery: Next Generation.* PAS Report 576. American Planning Association. 92–119. https://www.planning.org/pas/reports/pdf/PAS_576.pdf.

Ke, Y.C. (2010). One year after the Disaster (4): The Right to Return Home Cannot be Deprived. 88News.org. http://www.88news.org/?p=8518. [柯亞璇 (2010) 災後一年(4) 回家的權利是不可以被剝奪的]. (Accessed: 1 November 2014).

King, Ro. O. (2005). Hurricae Katrina: Insurance Losses and National Capacities for Financing Disaster risk. CPS Report for Congress: Received through CRS Web.

Knight, C.L., Burgin, G. and Chen, Y.F. (2013). Living with Typhoons: Disaster Management in Rural Taiwan. Final Report to the Royal Geographical Society (with IBG).

Kron, W. (2008). Flood Insurance: From Clients to Global Financial Markets. *4th international symposium on flood defence*.

Lee, H.C. (2010). Reformation of Taiwanese indigenousness and transmigration to China. *Alternatives*, 35, 241–258.

Lu, J.C., Chen, S.Y., Zhang, Z.R., Li, Y.N. and Zhuang, M.R. (2011). *Analyses of Recovery Issues and Policies after Typhoon Morakot*. NCDR 99-T36. National Science and Technology Centre for Disaster Reduction, New Taipei City. [盧鏡臣 (2011) 我國重大天然災害災後重建模式之研究-以九二一震災及莫拉克風災為例RDEC-RES-101-014 行政院研究發展考核委員會委託研究].

Lyons, M. and Shilderman, T., eds. (2010). *Building Back Better: People Centred Reconstruction.* Practical Action Publishing, Rugby, UK.

Maa, S. (2011). Disaster Management Framework in Taiwan Now and the Future Development. http://www.aec.gov.tw/webpage/control/emergency/files/train_4 99_3.pdf.

Mechler, R., Linnerooth-Bayer, J. and Peppiatt, D. (2006). Disaster insurance for the Poor? A Review of Microinsurance for Natural Disaster Risks in Developing Countries. ProVention Consortium. http://www.microfinancegateway.org/sites/default/files/mfg-en-paper-disaster-insurance-for-the-poor-a-review-of-microinsurance-for-natural-disaster-risks-in-developing-countries-draft-1–2006_0.pdf.

Ministry of Foreign Affairs. (2015). Info Taiwan. http://www.taiwan.gov.tw/sp.asp?xdurl=subject/Newsubject.asp&id=1&mp=1&ctNode=3474&subjectId1=43&subjectId2=50.

Ministry of Interior. (2014). Statistical Yearbook of Interior 2014. http://sowf.moi.gov.tw/stat/year/list.htm.

Morakot Post-Disaster Reconstruction Council. (2011). *Rebuilding a Sustainable Homeland with Innovation and United Efforts*. Morakot Post-Disaster Reconstruction Council, Kaohsiung.

Morakot Post-Disaster Reconstruction Council. (2012a). *Post Morakot Disaster Relief and Settlement*. Morakot Post-Disaster Reconstruction Council, Kaohsiung. [行政院莫拉克颱風災後重建推動委員會(2012a)莫拉克颱風災後救助與安置。].

Morakot Post-Disaster Reconstruction Council. (2012b). *Innovative, Effective Reconstruction: A Case Study in Da-Ai Community in Kaohsiung*. Morakot Post-Disaster Reconstruction Council, Kaohsiung. [行政院莫拉克颱風災後重建推動委員會 (2012b) 創新、效率的家園重建新思維：高雄山林大愛園區。].

Morakot Post-Disaster Reconstruction Council. (2014). *Rebuilding a Sustainable Homeland with Innovation and United Efforts*. 2nd ed. Morakot Post-Disaster Reconstruction Council, Kaohsiung City.

Perry, R.W. and Lindell, M.K. (1997). Principles for managing community relocation as a hazard mitigation measure. *Journal of Contingencies and Crisis Management*, 5(1), 49–59.

Professional Advisory Committee of Disaster Reduction, Executive Yuan. (2010). *Analysis and Policy Recommendation for Lessons Identified from the Morakot Disaster*. Professional Advisory Committee of Disaster Reduction, New Taipei City. http://morakotrecord.nstm.gov.tw/88flood.www.gov.tw/files/committee_plan0/116.pdf. [行政院災害防救專家諮詢委員會(2010)莫拉克颱風災害課題分析與政策建議]. (Accessed: 4 November 2014).

Quan, K.C. (2010). An investigation of the differences and conflict of reconstruction strategies between the government and aboriginals: From the view point of aboriginals. *Community Development Quarterly*, 131, 213–232. [全國成 (2010)以原住民的重建需求為觀點 – 探討家園重建政策與原鄉期待的落差與衝突。社區發展季刊，第131期，213–232頁。].

Quarantelli, E.L. (1995). Patterns of shelter and housing in US disasters. *Disaster Prevention and Management*, 4(3), 43–53.

Sa, Z.P. (2002). The Introduction of the Disaster Management in Taiwan. In 921 Earthquake Reconstruction Site, Emergency Operation Centre, Administration Yuan Disaster Education Training Programme, National Cheng Kung University Disaster Prevention Research Center. http://210.59.250.229/TNDS/Data/School/School_06.pdf. [薩支平 (2002) 我國防救災體系介紹。].

Shi, Z.F. (2002). The reconstruction of the history of Taiwanese aboriginals. In Shi, Z.F., Xue, S.K. and Dali, B. (eds.), *From Reconciliation to Autonomy: The Reconstruction of the History of Taiwanese Aboriginals*. Avaguard, Taipei. 305–332. [施正鋒(2002) 台灣原住民歷史重建。施正鋒、許世楷、布興・大立編，從和解到自治──台灣原住民歷史重建。頁305–32。台北：前衛出版社。]. http://faculty.ndhu.edu.tw/~cfshih/conference-papers/20020302.htm#_top (Accessed: 6 November 2014).

Shieh, J.C., Chen, J.S. and Lin, W. (2013). Skip to permanence without transition? Policymaking in Post-Morakot Reconstruction. *Taiwan: A Radical Quarterly in Social Studies*, 93, 49–86.

Sphere Project. (2011). *The Sphere handbook: Humanitarian Charter and Minimum Standards in Humanitarian Response*. Practical Action Publishing, Rugby.

Tsai, J.S. and Chi, C.S.F. (2010). Dysfunction of Governmental Emergency Systems for Natural Disaster Management – A Taiwanese Case Study. *Working paper proceedings: Engineering project organisations conference*. South Lake Tahoe, CA. http://academiceventplanner.com/EPOC2010/Papers/EPOC_2010_TsaiChi.pdf.

Tsou, C.Y., Feng, Z.Y. and Chigira, M. (2011). Catastrophic landslide induced by Typhoon Morakot, Shiaolin, Taiwan. *Geomorphology*, 127(3–4), 166–178.

Twigg, J. (2015). *Disaster Risk Reduction: Good Practice Review 9*. New ed. Humanitarian Policy Group, Overseas Develop Institute. http://goodpracticereview.org/9/.

UNDRO. (1982). *Shelter after Disaster: Guidelines for Assistance*. UNDRO, New York.

UNISDR. (2015). Global Risk Assessment Report on Disaster Risk Reduction 2015. http://www.preventionweb.net/english/hyogo/gar/2015/en/gar-pdf/GAR2015_EN.pdf.

United Nations Framework Convention on Climate Change (UNFCCC). (2007). Climate Change: Impacts, Vulnerabilities and Adaptation in Developing Countries. https://unfccc.int/ resource/docs/publications/impacts.pdf.

Water Resources Agency, Ministry of Economic Affairs. (2012). Flood Characteristics in Taiwan. http://eng.wra.gov.tw/lp.asp?CtNode=2299&CtUnit=619&BaseDSD=7.

World Bank Group. (2014). *Recovery and Reconstruction Planning in the Aftermath of Typhoon Haiyan (Yolanda): Summary of Knowledge Briefs*. The International Bank for Reconstruction and Development, USA. http://reliefweb.int/sites/reliefweb.int/files/resources/Recovery%20and%20Reconstruction%20Planning%20in%20the%20After-math%20of%20Typhoon%20Haiyan.compressed.pdf.

Wu, J.Y. and Lindell, M.K. (2004). A comparative study of housing reconstruction after two major earthquakes: The 1999 Northbridge earthquake in the United States and the 1999 Chi-chi earthquake in Taiwan. *Disasters*, 28(1), 63–81.

Wu, K.S. (1998). The System of Disaster Management in Serzi Island. MA Dissertation. Department of Public Policy, National Chung Hsing University, Taipei. [吳光雄(1998)我國災害防救體系之研究－以社子島防颱執行過程為例。國立中興大學公共政策研究所，碩士論文。].

Wu, Y.S., Hsu, H.H., Chou, C., Lu, M.M., Chen, C.T. and Chen, Y.M. (2013). An introduction to climate change in Taiwan: Scientific report 2011. *Global Aspect*, 3(1), 16–19. http://www.apectyphoon.org/sdt175/img/img/3858/Newsletter_March_2013/An_%20 In troduction_to_Climate_Change_in_Taiwan_Scientific_Report_2011 . . . p,16–19.pdf.

Xiao, H.Z. (2012). The development of aboriginal's land under New Liberalism: A investigation of the new partnership relation. *Cultural Studies Bimonthly*, 132, 44–68. http://csat.org.tw/journal/Content.asp?Period=132&JC_ID=620. [蕭惠中 (2012) 新自由主義下原住民族土地的發展地景：一個新夥伴關係的初探。文化研究雙月報。第132期，44–68頁。].

Zhang, Y. and Peacock, W.G. (2010). Planning for housing recovery? Lessons learned from Hurricane Andrew. *Journal of the American Planning Association*, 76(1), 5–24.

3 Human amplification of climate hazards

2010 floods in Pakistan

Inam-ur-Rahim and Henri Rueff

The 2010 floods in Pakistan were one of the greatest disasters in recent history (Gaurav et al. 2011). The floods impacted 55 per cent of Pakistan districts, washed away bridges, roads, public buildings and other infrastructure and inundated 2.4 million hectares of arable land (Brooker 2011). Over 1.6 million homes, 430 health facilities and 10 thousand schools were damaged in particular in Sindh province (WFP 2010). When the floodwater receded, approximately 14 million people (8 per cent of the country's population) were in dire need of humanitarian assistance (Brooker 2011). Remote sensing observations revealed 141 major embankments breaches (Ahmed 2011).

Unprecedented heavy monsoon rains triggered the 2010 floods in parts of the Indus River drainage area. The monsoon rains started on July 22, 2010 in the Northern Mountains of Pakistan and peaked between July 27–30 (Siddiqui 2010). The Indus' western tributaries catchment experienced record high rainfall reaching 450mm during those four days (Hashmi et al. 2012). Although the abnormal rainfalls have been widely documented, these climatic factors explain only partly the scale of the disaster. We argue that human factors at pre- and post-disaster levels explain the scale of the disaster.

This chapter describes first the historical and current facts about the Indus River basin. It then reviews the effect of variables that impacted the floods devastation. These variables are presented in the following order: climate, upstream unsustainable forms of land uses, land policy, power relation among landlords having different holding sizes, ill-adapted relief programmes and their incapacity to deliver assistance in most areas as well as various political perspectives which affected the above variables. In this way, the chapter examines the contributory role of human factors in climate hazards to improve climate hazard management strategies in Indus River basin

Taming the Indus River water for agriculture and energy

The Indus constitutes a large river system 3,180km in length with a drainage area of 960,000km^2 and an average annual discharge of 7,610m^3 s^{-1} with a catchment area of 399,000km^2 (Asif, et al., 2007). The river is fed by tributaries from the Himalayas, Karakorum and Hindu Kush (Riaz-ul-Haq 1994). The glacial cover of the

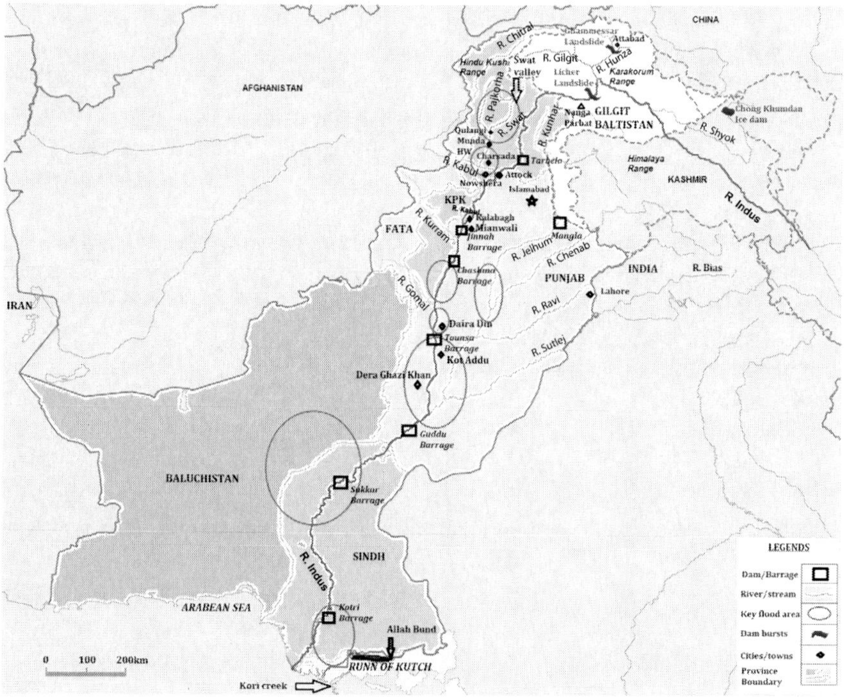

Figure 3.1 Indus River and tributaries. Circles identify key impact points

Source: Base map from Wikimedia commons, names and symbols sketched by authors

Karakorum (22 per cent) and Himalaya-Hindu Kush (8–12 per cent) regulates the Indus water flow through melting and restocking (Kreutzmann 2006). Upstream mountains and foothills receive annual precipitation from 760 to 1,270mm and contribute 72 per cent of the mean annual flow of the Indus River system (Tariq and Van De Giesen 2012). The rest of Pakistan receives less than 250mm annual rainfall which contributes 28 per cent to the Indus flow (Khan 2006).

Historically, the average coastal discharge of the Indus River was 106,000 cusecs (cubic feet per second). However, diversions for irrigation reduced this discharge to 28,000 cusecs with long periods without any flow (Asif et al. 2007). Out of the 140 mean annual water flow (MAF) in the Indus River system, 84 per cent flow is in summer and 16 per cent in winter. Of the 140 MAF, 18 can be stored in the large reservoirs built in Pakistan (Hanif 2003). The first diversion for irrigation was recorded during 1605–1627, when an 80km-long canal was constructed by the Mughal rulers to bring water from the Ravi River to gardens near Lahore. Work on the current irrigation system started in the 19th century under the British administration (Ahmad 2000). Before the subcontinent was partitioned, there was only the Sukkur barrage[1] built in 1932 (Magsi and Atif 2012).

After the partition, other barrages were built including Kotri in 1955, Tounsa in 1958 and Gudu Barrage in 1962. The present irrigation system also contains three storage reservoirs at Mangla, Chashma and Tarbela that became operational in 1967, 1971 and 1976 respectively (Mohtadullah et al. 1992). Until 2011, 19 barrages, 43 canal systems with 48 off-takes were built (Magsi and Atif 2012). About 22.05 million hectares of the Indus basin are classified as cultivated area and the canal irrigation system is making cultivation possible over large swathes of land (Naseer 2013).

Today the Indus River consists of a large contiguous and man-made system of 61,000 km of canals and 105,000 watercourses irrigating about 14 million hectare of land (Magsi and Atif 2012). So far 6,807km of embankments and 1,410 spurs have been constructed to tame the Indus River system. At partition in 1947, the boundary line between India and Pakistan was drawn on religious grounds keeping 85 per cent of the irrigated area of the Punjab in Pakistan, while the part of Kashmir containing the headwaters of the Punjab Rivers remained in India (Bisht 2013). The high population growth in Pakistan increased the need for food production and reduced the per capita water availability from more than 5,000m^3 at the time of independence to 1,000m^3 compared to the global average of 8,200m^3 (Imam and Lohani 2012). Water distribution disputes are increasing both between the two countries, with India building dams on various rivers flowing into Pakistan (Noor-ul-Haq 2010) and on water shares between the provinces within Pakistan (Imam and Lohani 2012).

Indus River basin floods in historical context

Flooding and drastic changes of the river courses are believed to be the main instrumental factors in the collapse of ancient Indus valley civilizations that comprised nearly a thousand settlements, including Mohenjo-Daro and Harappa, during their peak at around 2600 BC (Atta-ur-Rehman et al. 2013). Further studies of Buddhist archaeological remains along the Swat River (Indus River tributary) provide accounts of various visiting pilgrims between 400–628 AD that attest to frequent flooding of the River Swat (Rahim and Viaro 2002). The famous Buddhist stupa of Gulkada, excavated near Mingora, reveals several layers of reconstruction after recurrent flooding.

Floods in the eastern and western tributaries of the Indus basin usually do not coincide due to topographic divisions between the various mountain ranges (Riaz-ul-Haq 1994). The high rainfall during the monsoon season (July–September) in the foothills of the Himalaya often causes additional flooding in the eastern Indus tributaries (Inam et al. 2007). As such, the Indus basin has witnessed high floods in 1973, 1975, 1976, 1978, 1986, 1988 and 1992, primarily originating from the eastern tributaries (Memon 2011). Topological changes from earthquakes and erosion, as those found in the plains of Punjab and Sindh, generate changes in the river course. One such change occurred as a result of the 1819 earthquake, when a 5,120km^2 area was submerged for two years by a 3–8m naturally raised ridge called "*Allah Bund*" (see Figure 3.1). Later, in 1826, the Indus floods eroded a

passage through an old channel to discharge into Khori creek close to Runn of Kutch (Shroder Jr. 2004).

Major floods also result from breached dams and landslides; the 1929 and 2010 floods (ADB and WB 2010) were direct outcomes of dam bursts. It is recorded that during the 2010 flood, the peak Indus River discharge was observed at Tarbella dam level at 833,000 cubic feet per second (cusecs) (FFC 2012). When in 1929, the Chong Khumdan ice-dam,[2] located in the Shyok valley of the Karakoram Range, ruptured, its discharge reached an estimated peak flow rate of 689,000 cusec and produced a wall of water over 15m high at Attock (Cornwell and Hamidullah 1992). This in turn blocked the flow of the adjoining Kabul River and inundated large tracts of land and settlements around Nowshera in the Peshawar Valley. Another example occurred in 1841 as the Lichar landslide on the west flank of Nanga Parbat blocked the Indus River and resulted in a dam-burst flood wave over 28m high at Attock (Cornwell and Hamidullah 1992). This was repeated again in 1858, when another massive slope failure (Ghammessar landslide, 125 million m^3) took place just downstream of Attabad on the River Hunza, resulting in a 20m high flood wave at Attock (Schneider et al. 2011). Over 36 such floods have been documented in the Indus valley since the late 18th century alone (Cornwell and Hamidullah 1992).

Climatic factors contributing to 2010 Indus floods

Global warming and climate change have been termed as the main causes of increasing extreme events in the region. Temperatures have increased at the rate of 0.057°C per decade in the last century in Pakistan (Chaudhary et al. 2009). Mean temperatures have risen at the rate 0.099°C per decade from 1960 to 2007, resulting in total change of 0.47°C (Chaudhary et al. 2009). The 2001–2010 decade was the warmest decade recorded globally; 2010 ranked as the warmest year (+0.53°C) (WMO 2012). The Karakorum and Hindu Kush mountain ranges are generally winter precipitation zones; where the precipitation at 1500–3000m altitude scarcely exceeds 130mm $year^{-1}$, while at 4,000–6,000m the precipitation levels varies between 500–1000mm, making these a highly glaciated zone (Iturrizaga 2008). During winter and spring, the Karakoram and Hindu Kush Mountains are affected by broad-scale weather-systems, originating primarily from the Mediterranean or Caspian Sea (Syvitski and Brakenridge 2013). Some argue that due to global warming, frozen water resources in the mountain ranges have been losing their reserves at an unprecedented rate and that out of 2,500 reported glacial lakes, 52 are vulnerable to glacial lake outburst floods (Rasul et al. 2011). On the other hand, some researchers have noticed glaciers expansion at relatively higher elevations and the glacial melt at lower elevation in the Karakorum referred to as the Karakorum anomaly (Hewitt 2005).

Currently, most climate studies of the area focus on the greater Himalaya glaciers; limited attention has been paid to climate change in the trans-Himalayan region of the Hindu Kush and Karakorum ranges (Archer and Fowler 2004). This is unfortunate, as monsoon floods that originate from the western tributaries,

particularly in the Hindu Kush Mountains, are a rare phenomenon (Houze et al. 2011). Usually, this type of rainstorm produces copious monsoon rain far to the eastern part of Himalaya, over the mountains and wetlands of northeastern India and Bangladesh (Houze et al. 2011). The rainstorms responsible for the 2010 floods originated from a watershed region located in the western tributaries of Indus (Houze et al. 2011). This unprecedented monsoon of 2010 in Hindu Kush Mountains was further intensified by the development of a low pressure zone over northern Pakistan (Atta-ur-Rahman and Khan 2013).

Some argue that monsoonal zone of Pakistan (receiving 65 per cent of monsoon rains) has shifted 80–100 km from northeast (upper Punjab and Kashmir regions) towards Northwestern Khyber Pakhtunkhwa and Punjab regions (Khan 2011). By comparison, southern Sindh Province, which suffered the worst flooding, received little rainfall throughout the 2010 summer monsoon (Syvitski and Brakenridge 2013). Other researchers view the monsoon circulation pattern of July 2010 as abnormal and part of a long-term trend that defies the typical monsoon dynamics expected in northern Pakistan (Wang 2011). However, during the last 31 years, there is no evidence suggesting that the northward extension of the monsoon has increased (Wang et al. 2011a).

The heavy rains of 2010 are thought to have been initially formulated by a combination of a jet stream of strong winds high in the atmosphere over northern Eurasia together with *La Niña* which induced high sea surface temperatures with large amounts of moist air over the Indian Ocean (Oxley 2011). Specifically, *La Niña* created monsoon depressions that carried sufficient moisture towards northwest leading to extensive rainstorms and flooding (Khandekar 2010). The depressions blocked the warm and saturated air that normally moves from west to east. This same ridge prevented the rains from reaching western Russia where a record severe drought and hot summer are blamed for wildfires during 2010 (Gronewold 2010). The two record-setting extreme events during 2010 summer appear to be mutually connected (Lau and Kim 2012) as a manifestation of climate variability or are connected in some measure with global climate change (Gronewold 2010).

Anthropogenic factors contributing to the flood disaster

Since the 1940s increasing anthropogenic factors have contributed to climate change. This includes increased industrial emissions, exploding population growth, development and changing lifestyles that have raised human influence on climate change (Rasul et al. 2011). The growing population has triggered harmful land use practices in the absence of appropriate land use policy. This section highlights poor and inappropriate land use in different ecological zones and corresponding policies that further affected the 2010 flood impact. Other factors are also considered, such as topographic variations; whereby flash floods waters receded within days in mountain regions, in comparison to where water receded only after weeks and months in the plains of Punjab and Sindh (Kirsch et al. 2012)

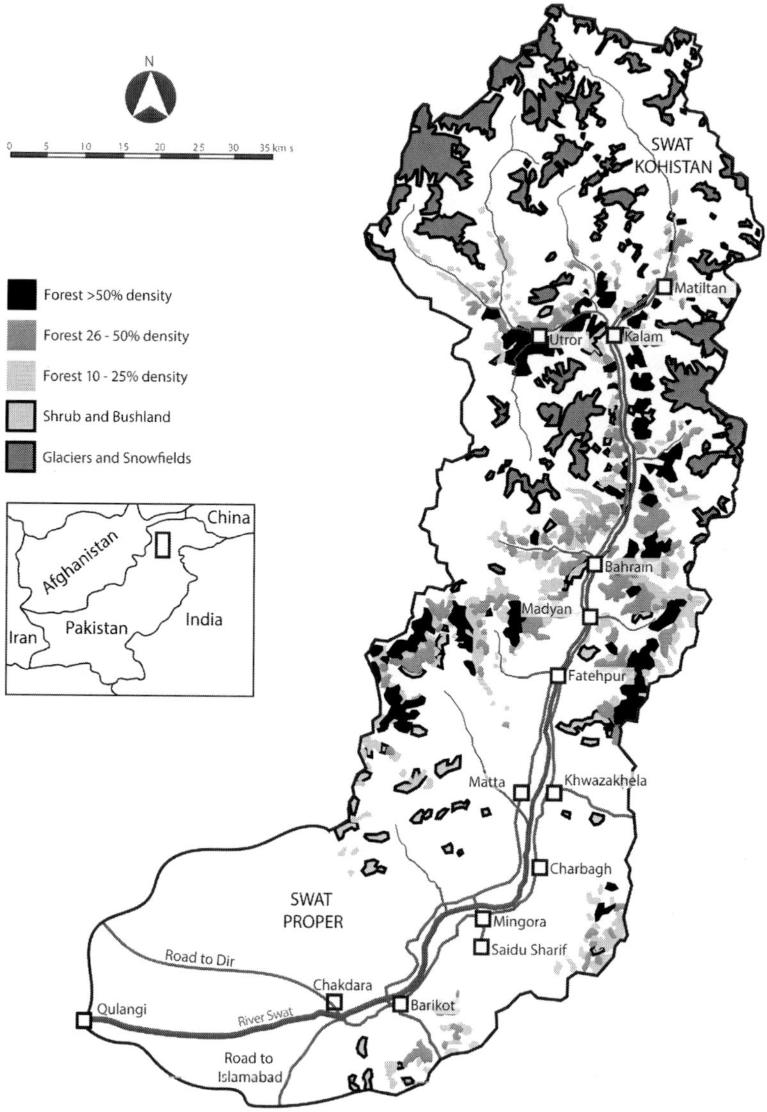

Figure 3.2 Swat valley map

Source: Steimann 2003

Inappropriate land uses in the mountain regions

As one of the sources of the 2010 floods in the mountain region, we will focus on the case of the Swat valley; located in northeast Pakistan. The normal average discharge of the River Swat, at Chakdara, during the month of August, is 10,900

cusecs (PDMA 2014). During the 2010 flood, the peak discharge at Chakdara reached at 259,000 cusecs, 100,000 cusecs greater than the 1929 floods. Downstream, after receiving water from Panjkorha River originating from Dir valley, the discharge of the River Swat increased to 367,000 cusecs before entering Charsadda and Nowshera plains. At Nowshera the river discharge increased to 500,000 cusecs after merging with River Kabul (PDMA 2012).

The Swat valley is generally divided into Swat Kohistan (north) and Swat proper (south) where it is wider. Swat Kohistan is composed of highly elevated valleys and mountains (up to 6,000 m) and serves as the source of the River Swat, a western tributary of Indus River. The River Swat flows from northeast to southwest through the middle of the valley formed by the erosion of the River Swat and its tributaries (Rahim and Viaro 2002). Variations in the flow velocities produce diverse sediment load capacities in Swat Kohistan and Swat proper (Cornwell and Hamidullah 1992). Usually boulders are deposited in Swat Kohistan, where the catchments have higher flood velocities; these are followed by cobbles, pebbles, granules, sand, silts and clay deposition with reducing velocities in Swat proper (Kirsch et al. 2012).

The flash floods of 2010 are the first of their kind in local population's historical memory. The recorded average rainfall during four days from July 27–30 in 2010 was more than 350 mm compared to an average of less than 100 mm during the entire month of July (Hashmi et al. 2012). The pattern of the rainstorms was also unique. For instance, in some sub-valleys, like Utrore, Daral and Chail valley (25–40km^2 each), the intensity of rainfall was very high with multiple cloudbursts. In other sub-valleys, like Gabral, Ushu and Miandam, the intensity was relatively low and did not generate flash floods. Topographical features such as elevation, valley orientation, aspect and slope as well as height and number of upwind barriers to the airflow, might explain these occurring anomalies (Syvitski and Brakenridge 2013). Other factors are also taken into account as contributing to the 2010 flood's disastrous consequences; these include illegal timber logging (Lewis 2010), encroachment on riparian area (Hashmi et al. 2012), overgrazing and the use of steep slopes for cultivation (Hussain 2010).

Ambiguous tenure, deforestation and steep slopes cropping

Land tenure of forest resources is unclear. After the merger of the princely Swat state into Pakistan in 1969 (Rahim and Viaro 2002), the provincial government declared the Swat and Dir Valley natural forests as protected. Once protected, these forests fall under the jurisdiction of forest law, are property of the government and are not to be used for any form of cultivation. The government is the *de jure* owner and manager of all non-cultivated lands, including forests and pastures. However, because of the lack of land use and resource harvest regulations, the *de facto* ownership actually rests with the local forest-fringe communities.

Landowners have carried out *de facto* appropriation of different natural forest unit on clan and family basis. They are gradually clearing the forest trees for their timber and fuel needs and using the newly-cleared land for off-season vegetable

cultivation of potato and peas (Rahim and Viaro 2002). The off-season vegetable production in Swat Kohistan was initially introduced in the early 1980s to improve the economic conditions of the communities and reduce their dependency on depleting forest resources. This resulted in a 4.2–5.0 per cent annual deforestation rate in the Swat Kohistan over the last four decades (Khan 2010).

Steep-slope crop cultivation is practiced in dry cold conditions and experiences high soil erosion that quickly renders the land unproductive and consequently abandoned. According to our data, the income from off-season vegetable cultivation is 5–10 times higher than prevailing livestock production. The continued choice to mono-crop further contributes to soil fertility loss, lowers production and the increased need for chemical fertilizers and pesticides, which in turn, leads to reduced economic returns (Rahim 2010). Instead of constructing proper terraces on cultivated hill slopes and crop rotation, which promotes land productivity and sustainability, farm labour is used to clear forests and bring virgin hill slopes under cultivation. Unclear tenure coupled with the lack of regulations for land and resource use leads to such activities and contributes to ecosystem vulnerability (Rahim 2010). To avoid losses, new lands on steeper slopes are carved out from the natural forest and pasture areas every year. According to our data, the cash crop cultivation above 33° slopes had reached 9 per cent in Burhawai valley of Naran uplands through 2010.

Trees from the cleared forests in mountain ravines that were cut due to unregulated land use were dislodged by 2010 floodwaters destroyed downstream bridges and roads. This exacerbated destruction in flood-devastated areas and reduced accessibility to the damaged structures intended for flood protection (Lewis 2010). Consequently, deforested and tilled lands on steep slopes were more susceptible to flooding, landslides and soil erosion and directly contributed to the volume and viscosity of the floods.

Encroachment over the riparian area for tourism

Until 1920, all land tenure in the Swat valley was communal; settlements were strategically located on raised grounds with limited cropping possibilities. The valley was isolated with all the agriculture produce designated for local subsistence (Rahim and Viaro 2002). The riparian area was covered with successfully adapted indigenous shrubs and trees (Nilsson and Berggren 2000) alongside local food crops of rice, wheat, barley and orchards. After 1920, with the gradual abolishment of communal land tenure together with the development of accessible road network development, Swat valley became a successful tourist hub (Rahim and Viaro 2002).

An example of changing land use in Swat valleys is that the riverbanks have become prime sites for hotel construction. Due to their commercial importance, the land along the riverbanks is three to four times more expensive than other such lands (Rahim and Viaro 2002). To maximise land availability, a small plot on the riverside is purchased and approximately twice the area of the purchased land is reclaimed from the riverbed by erecting pillars or a retaining wall. Thus, the encroachment on the riverbed by hotels has restricted the flow space for floodwater.

Trapped in the narrowed passage, large size boulders and tree logs that were swept from ravines caused most of the damage in the newly developed tourist areas along the river. The logs dammed the river flow below concrete bridges for several hours, resulting in momentary dam bursts that amplified the level of devastation further along the valley. As a consequence, the abrupt reduced water velocity filled the riverbed with boulders, cobbles and pebbles. This raised the riverbed by 4–5m, submerging some of the riverside hotels, bridges and settlements. This further diverted the floodwater to submerge croplands up to 200–300m away from the regular riverbed.

The impact on the urban sector was felt along the two tributary streams of the River Swat that pass through Mingora, the main city of the valley. In the absence of a proper drainage system, the streams serve as the main water conduit in the city. As land by the streams is cheaper, the poor residents are the main homeowners. These settlements are highly exposed to floods, and, during the 2010 flood, almost 2,000 houses in the area were destroyed.

Exotic plants in riparian areas

The local riparian zones of the Swat valley possess an unusually diverse array of plant species; adapted to variable flood regimes, geographically unique channel processes, altitudinal climatic shifts and upland influences on the fluvial corridors (Naiman and Décamps 1997). The plant species sustainability along the river depends on the interaction between erosion and sedimentation on the river margins (Nilsson and Berggren 2000). Until the abolishment of communal land tenure, the riparian areas contained numerous native shrubs and trees. For example, indigenous Salix trees were commonly found on the riverbanks and aits in the Swat Valley. Salix trees are specialised to riparian areas because their stem flexibility can endure and resist potentially high levels of the shear stress accompanying seasonal floods (Featherston et al. 1995). Salix also have the ability to root in barren and unstable soils, thus making it the most valuable tree for erosion control (Margolin 1986). Hence, the Salix and other native riparian vegetation actively reduce the force of floodwater and its harmful effects (Cohen 2014).

Poplar (*Populus deltoides*) is an exotic species introduced in Pakistan during the late 1950s. It became popular due to its economic benefits as a commercial timber tree and its fast rate of growth (Khan and Choudhry 2007). In turn, the local people started to clear the native trees and shrubs in the riparian area and aits[3] of the River Swat and to replace them with the fast-growing, shallow-rooted trees. These newly introduced trees were also planted on field boundaries and were uprooted during the 2010 floods. According to our data, tens of thousands of uprooted poplar trees were deposited in the plains of Peshawar valley during the 2010 flood.

Dumping of solid wastes on streamside and riverside

Between 1970 and 2010, the population of Mingora, the main city of the valley, increased at least tenfold (from 27,000 to 290,000), but the drainage and solid

waste management systems did not develop in line with the city growth. Approximately 70 tons of solid waste is generated daily in Mingora city (Rahim and Viaro 2002). In the absence of landfill and solid waste recycling systems, local streams and the River Swat became the prime sites for dumping municipality collected solid waste.

In locations where solid wastes cannot be collected by the municipality, small dumping sites on vacant plots are usually used. In the past, the semi-composted waste was transported on mule back to croplands as manure. But now, with increased use of plastic that make a significant part of the solid waste, it can no longer be used and is left to accumulate. During heavy monsoon rains, solid waste that had been dumped was washed away by stream torrents, contributing to the volume of the flood (Rahim 2010). Inadequate solid waste management together with economic-driven land use directly contributed to destructive impact of the 2010 floods including water volume, blocking drainage and resultant health issues.

Contributing factors in the plains of Punjab and Sindh

The suspended sediment concentrations in the Indus River system vary from less than 100 ppm (parts per million) at low flows to over 5,000 ppm at high flows (Siddiqui 2010). The sediment fluxes in the Indus river system are substantially higher than other rivers with mountain sources around the world (Collins 1996). In the mountain valleys, floodwaters receded within days compared to the several weeks it took in Punjab plains, or the months in some areas of Sindh (Kirsch et al. 2012). Consequently, the varying velocities and flow height formed different impacts in different regions; for example, a reduction in flow velocity increased deposits of sediment load (Cornwell and Hamidullah 1992). The River Swat's high sediment load together with its relatively extreme elevation have set the stage for levee failures and channel avulsions. The planning of major water and sediment flow diversions of future floods can be instrumental for effective future flood control along the Indus; as neither an improved flood warning system nor the reinforcement of existing engineering structures can avoid future catastrophic flood impacts (Syvitski and Brakenridge 2013).

Engineering structures for irrigation and flood control

In the past, flood plains along the Indus River had a very dynamic flow pattern; however, now it is constrained by embankments on both sides and by several constructed barrages (Gaurav et al. 2011). The effectiveness of these current river control strategies on the river's high sediment loads is compromised by a lack of space and thus constricted water flow (McGuire 2012). The construction of dams and barrages for storing water for irrigation is leading to rapid aggradations, particularly upstream. This in turn has caused a significant increase in cross-valley gradient, leading to breaches upstream of the barrages and inundation of large areas (Gaurav et al. 2011). For instance, the original capacity of the three big regional reservoirs of 18 MAF, has already been reduced by about 4 MAF through

sedimentation (Hanif 2003). The sediment delta has expanded to 8 kilometres from the dam whilst the riverbed level has increased 61 m at the pivot point (Izhar-ul-Haq 2014). Damage caused by dam and barrage-related backwater effects[4] has reduced water flow and sediment conveyance capacity and led to multiple failures of irrigation system levees.

Encroachment of riparian area

Before the Tarbela Dam was operational in 1976, Sindh would experience a yearly 300,000–500,000 cusecs flood. Over centuries, flood patterns have shaped the river and corresponding social and administrative systems. However, in the post-Tarbela years reduced incidence of low or medium-level floods exposed vast areas of riparian forests and marshes (locally known as *katcho* land) for human settlements and agriculture (Siddiqui 2010). Since the construction of the Tarbela Dam, weak regulatory measures allowed, approximately, 500,000 acres of *katcho* lands to come under human settlement in Sindh. The Sindh landlords have constructed illegal bunds (locally known as *Keiti*) to protect the grabbed *katcho* land from floods. This has reduced the waterways of the river, resulting in higher floodwater levels against the bunds (Siddiqui 2010). Other structures, such as bridges and barrages, have choked flood plains with obstacles that interfere with the natural streamflow. Illegal local dykes, constructed to protect agricultural land on the flood plains, have also disturbed the river and caused it to swell with high waves near flood-protection embankments (Memon 2011).

Anthropogenic factors amplifying the impact of flood disaster

Poverty, lack of resources and low adaptive capacity to climate change by the local population in the Indus Delta are exacerbating the vulnerabilities and challenges to sustainable food production. (Rasul et al. 2012). For example, overflow due to the raised riverbed submerged large tracts of agricultural lands in Dera Ghazi Khan where the flood engulfed about 23 per cent of the cultivated area and damaging over 17,600 acres (8 per cent) of cotton crops and 16,476 acres (7 per cent) of rice paddies. The harder-hit district of Mianwali suffered a deluge of an estimated 93 per cent of its cultivated land (790,000 acres) (Khan et al. 2010).

Impact amplifying factors in mountain valleys

Disaster impacts and the role of local actors

The 2010 floods affected 90,500 residents in the Swat district alone as all 45 bridges on the River Swat and its tributaries were swept away (PDMA 2012). A news report accounts how the 2010 floods washed away more than 14 basic health units, 1,575 commercial units (including hotels and shops), 70 km of road and 26 schools (Nation 2010). According to our estimates, 6,800 land-owning families in

Swat Kohistan lost part of their terraced croplands during the 2010 flash floods that washed away 43 per cent of the cultivable terraces. In the foothills and low-elevation valleys, the key impact on farmland was erosion (Rahim 2010). The 2010 flood damaged about 400 hectares of commercial crops, 350 hectares of domestic farms and about 125 hectares of linear plantations[5] in lower Dir district (Khan et al. 2012).

There is no flood warning system operating in Swat Kohistan, but fortunately, the flood occurred during the day, which minimised the loss of life. The sector to suffer the greatest socio-economic impact of the 2010 floods were the wealthy as they lost their riverside lands, their hotels on its banks and their cultivated rice fields along its edges. In comparison, the middle class and poor social communities were mildly affected due to their habitation of the remote areas on mountain slopes away from the streams.

Presently, traditional social values are still partially in place, exemplified by shelter provided to local residents. The displaced landowners' class were accommodated by their relatives, and the affected tenant families were accommodated by their landlords. However, the transient labour force lost both their rented abodes and private belongings and found themselves without shelter and food immediately after the flood. Further hardships affected the community as prices increased for available accommodations and travel as people exploited the situation. In some localities, the available food stock vanished, creating famine-like conditions as shopkeepers sold the food items at very high prices.

Difficulties and biases in relief activities

The relief agencies were able to arrive in the Swat valley only five days after the disaster. Unfortunately, the relief activities were not properly organised and duplication occurred almost everywhere. As a consequence, multiple agencies were distributing the relief aid in the same place, often concentrated on roadsides and easily accessible areas. Most organisations failed to focus on remote areas, hence accessible communities were over-supported whilst remote areas were neglected. Moreover, active NGOs were allied to local CBOs[6] and provided aid based on favouritism rather than on merit. In clear contrast to other relief agencies, the better prepared military created a more realistic list of the affected families and provided special cards for the receipt of relief goods.

Due to the almost-flat topography in Punjab and Sindh, the main impact on the plains was the long duration (10–50 days) of inundation of cultivable lands and rural settlements. The stagnant floodwater kept the homes of poor rural people underwater, forcing them to move to raised areas and remain cut-off from the rest of the country and exposed to the caprice of the weather and water-borne diseases (Rahim 2010).

The lack of regular maintenance of river embankments, barrages and ill-planned structures along the river also contributed to the magnitude of the devastation (Iqtidar 2010). Remote sensing of the 2010 flood revealed 141 breaches in the river embankments across the country, 136 in the plains of Punjab and Sindh

(Ahmed 2011). In some areas, the mega-floods were created by the very structures that were constructed to contain them (Iqtidar 2010). Moreover, the barrages and dams reduced the frequency of seasonal floods, resulting in the increased salinisation of flood plains (Nilsson and Berggren 2000) affecting 5.9–6.3 million hectares (Mirza et al. 2005). The reduction meant intermittent flooding, which would reduce soil salinity and alkalinity, and did not occur (Ahmad 2011).

Strategies of relief and loss reduction

Pakistan is a signatory to the Hyogo Framework for Action (HFA), a 10-year disaster risk reduction plan established in 2005 that describes strategies aimed to reduce losses caused by disasters (e.g. lives, social, economic and environmental costs). The guidelines list key activities to build national resilience and outline five Priorities for Action (PFA). Pakistan became a signatory of the framework after the 2005 earthquake exposed vulnerabilities of the existing disaster risk management strategies adopted by the country. In 2006, the National Disaster Management Ordinance (NDMO) 2006 was introduced by the Government of Pakistan to provide a legal framework for disaster risk reduction at federal, provincial and district level and included 9 priority areas from the HFA. In 2010, the Hyogo Framework was enacted under the National Disaster Management Act 2010 with the intent that provincial chapters of the NDMA[7] would act as a central hub for implementing, coordinating and monitoring disaster management (NDMA 2011a). The purpose was to effectively coordinate the disaster management activities of the donor agencies, government agencies and non-government organisations (White 2011).

NDMA coordinated the relief operations carried out by the Pakistani government agencies, the army and international agencies such as the World Health Organization (WHO), the United Nations Children's Fund (UNICEF), Médecins Sans Frontières and the US Agency for International Development (USAID) (WHO 2010). The 2010 floods saw the establishment of five thousand relief camps by the government, 400 by secular civil society organisations and 29 by religious 'extremist' groups. NDMA has yet to progress its disaster risk management at sub-national and district level (White 2011). The inability to access information management systems by the District DMAs has led to incorrect assessments of damages. As a result, there were many districts that were only partially covered while others had excessive coverage. As highlighted, an inadequate monitoring and evaluation system was in place (NDMA 2011b).

In response to this disaster, the federal government developed a nationwide distribution programme for cash aid, based on smart cards called the *Watan Cards*. These cards are being distributed to flood survivors for withdrawal of relief funds from Automated Teller Machines (ATMs) (B4A 2010). However, corruption and malfeasances were commonly observed during the distribution process of Watan cards. Such actions occurred when the district revenue department officials engaged private agents to provide the Watan Card (PACF 2012). In some cases where Watan cards could not be provided, bank cheques were provided to the affected families through the district management officials. Our data shows that

in 35 per cent of the cases, the respondent reported that the respective banks did not honour the bank cheques given by the government agencies. There are also reports of bribes taken by the district revenue department when enlisting the victim into the intended beneficiary (PACF 2012).

Political factors contributing to the flood disaster

Ownership of irrigated land in the Indus Valley is concentrated in the hands of a few large landholders – 2 per cent of the landowners control more than 45 per cent of all land. This severely constrains agricultural competitiveness, livelihood opportunities and keeps a huge proportion of the rural population landless (Giampaoli and Aggarwal 2010). Between 20 to 40 per cent of rural households are reported to be landless or near-landless; they lease, sharecrop land or work as labourers for survival (USAID 2010). Landlords with large holdings are also powerful tribal leaders dominating the rural economy and society, particularly in rural Sindh and Southern Punjab (Budhani and Gazder 2011). These landlords and tribal leaders are, in most cases, the political elites influencing policies and practices with large impacts on the rural population.

Current irrigation management and flood control measures in the Indus plains are primarily aimed to save cities and villages with strong political connections. Powerful landlords, to the detriment of poor farmers, divert irrigation water well above their due share by using their political influence or by bribing irrigation officials (Azam and Rinaud 2000). Though regulations for the use of irrigation water do exist, it is politically influential farmers who determine what rules and rights are implemented (Mehari et al. 2007). Powerful landlords in Sindh illegally occupy the riverine forest lands (*Kachu*) for cultivation and have constructed bunds (*Keiti*) to protect these lands from floods. This practice has reduced the waterway of the river, resulting in increased floodwaters height against the bunds (Siddiqui 2010). Furthermore, there are no regulatory measures in place to prevent deliberate breaches in canals and embankments by powerful landowners, with support from irrigation authorities, which submerge the properties of powerless communities downstream (Gaurav et al. 2011).

In response to the devastation caused by the 2010 floods, cheap loans and other aide facilities were provided by the government for the victims. Unfortunately, these were diverted to the feudal lords and big farmers, with little offered to the poor peasants and small farmers (Looney 2012); causing ripple effects along the affected areas as the percentage of malnutrition in local households increased by more than 60 per cent (WFP 2010).

Furthermore, political representatives manipulated the distribution of the relief goods on the basis of political affiliation to strengthen their vote bank, rather than the extent of need. For example, there are reports that *Watan Card* funds were diverted from actual affected households by ruling political parties who channelled the funds to voters in their winning constituencies. Hence, those with political alliances were the prime recipient of relief instead the actual affected people. The current political system could not break the ethnic societal stratification and at

the same time increased the number of a politician's supporters. As such, even the social and economic services for the public were prioritised on political basis. In such a situation the current formal political system has lost its effectiveness in responding to disasters until proficient civil institutions are built free from such political interference.

Disaster management is better served when those who control where the funds go are kept rigorously separated from those who benefit from their misdirection. Greater transparency in both the process of entitlement to *Watan Cards* and use of massive amounts of personal data collected would be beneficial in combating violations (B4A 2010).

Vested interests of the political and feudal elite along with poor planning and incompetence amplified the scale of devastation on the plains (Iqtidar 2010). The absence of a viable flood warning system also intensified the impact of the 2010 flood disaster. Consequently, there was lack of time to collect animals and the grains; the sole savings of most villagers (Iqtidar 2010). The Flood Inquiry Commission, appointed by the Supreme Court of Pakistan, identifies the present early warning system that could have minimised losses or entirely avert this catastrophe as ineffective. Administrative failure of provincial irrigation departments, rampant corruption, criminal negligence and omnipresent encroachments in the flood plains were identified as key causes of the disaster in Sindh and Punjab (Memon 2014).

Discussion and conclusion

Countries in the Hindu Kush-Karakorum-Himalaya have been facing disaster challenges over millennia (Elalem and Pal 2015). Most of the mountain ranges are situated on fault lines (Joshi and Khan 2009) and a large number of earthquakes have jolted the region, including the 2015 earthquake in Nepal and the 2005 earthquake in Northern Pakistan (Hiroshi et al. 2007; Galetzka et al. 2015). Most of these mountains are fragile in nature (Jodha 2005) and experience glacial lake outbursts, high rates of erosion, particularly during the monsoon season (Rasul et al. 2011) which results in the silting of river beds, which causes rivers to change course and make the downstream plains more vulnerable to floods (Collins 1996). At the same time, droughts in arid regions are becoming more common (Anjum et al. 2012) as rising sea levels due to global warming (Meehl et al. 2005) threaten the coastal areas.

The frequency of extreme weather events in particular the floods, are likely to increase as the changing climate continues to pose major challenges. At the same time, human factors amplifying such disasters are ignored. The lack of investment in disaster-resilient infrastructure in disaster-prone areas then burdens local households. After the 2010 flood disaster, all the damaged public schools, health facilities and communication infrastructure were repaired and rebuilt through significant spending while affected people were left to endure the full costs of rebuilding their damaged homes. In addition, crop and livestock insurance was unavailable to help the affected population. The introduction of crop, livestock insurance schemes and the enforcement of the land use regulations can result in more effective disaster prevention and aide response.

Despite recurrent flood disasters, viable local level flood warning systems do not yet exist, highlighted by the fact that even with the availability of modern communication technology, loudspeakers are still used for flood warnings. Mobile phones and FM radios can be used more effectively to extend timely flood warning at the household level.

The NDMA reported a lack of awareness amongst the institutions and communities to integrate disaster risk reduction (DRR) as an essential part of sustainable development. The result, in effect, is that DRR will not be treated as a priority by state institutions and communities. Moreover, although implementation of DRR policies and strategies is crucial, lack of resources and dire poverty makes it difficult for the government to earmark substantial funds for DRR activities. The increasing dependency syndrome of local institutions and communities when dealing with local disasters is also a major challenge. This attitude will most likely prevail until the process of decentralising the responsibilities for disaster management is at local and community levels.

High population growth and corresponding pressure on land resources in mountain areas can be directly related to downstream floods. Despite the fact that the government has declared the non-cultivated land as government property, mountain inhabitants are continuously encroaching on natural forest and pastures on mountain slopes; due to the absence of appropriate land use and resource harvest regulations. A devised policy and corresponding regulatory measures for land use in riparian areas and mountain slopes complemented with land use planning for vulnerable localities along the river/stream beds will facilitate an improved resolution to current land use and decrease the fear of extreme events like the 2010 floods.

Under the current ad hoc approach, billions are lost in damages due to floods and corresponding rescue efforts, and relief and rehabilitation activities. A significant increase in the focus on disaster preparedness, risk reduction and making communities safer and more resilient would save lives and cut losses. For example, additional climate monitoring at the sources of the floods will aid in future warning systems. Currently, the density of measuring stations in mountain regions is much lower than in lowland areas; stations are generally concentrated at lower elevations in valleys and, thus, give a biased representation of the climate (Awan 2003). Additional weather and water discharge gaging stations in remote, high-elevation valleys can increase the precision of forecasting of extreme events that would help forestall the damages.

Management, rather than control, of floods in the Indus river basin with supporting measures can reduce the negative impacts on peoples and resources (MWC 2007). Such measures may include a change in engineering the response from the control of flood approach to that of a management response (Webster et al. 2011). For instance, as high sediment load will continue to raise the riverbed in relation to the surrounding land, it may become increasingly difficult to reduce future flood impact. The 2010 floods have revealed that periodically flushing the sediments in the channels can reduce the need for costly and damaging levee-breaching during flood events. Adapting to the rhythms of the Indus Basin

rivers, instead of manipulating the river, can further decrease any future flood impact on its surroundings. Provision of natural inundation zones and restoration of wetlands all along the Indus river and its tributaries could help in moderating high flood peaks; in addition to providing important ecosystem services such as groundwater recharge, carbon sequestration and maintaining biodiversity (Mustafa and Wrathall 2011). Mountain landscapes need more focus for effective flood disaster management.

The regulatory mechanisms for the provision of public services in Pakistan during rescue and relief, based on past experiences, are weak. The continued ad hoc relief is creating a dependency syndrome that restricts attempts for long-term adaptation strategy; such as ecosystem-based adaptation measures, which is considered one of the most overall and cost effective strategies to mitigate the impacts of the extreme events. Local economic and business activities are not regulated and thus exploit the flood-affected populace. Effective mechanisms that streamline and monitor these activities, without political interference, and that are integrated into the overall adaptation and disaster-risk education can protect flood victims from future harm. On the ground, NGOs are creating alternative social structures to the traditional in the form of CBOs that may contribute to the problem rather than part of the solution. To ensure their success, the newly formed social structures could be based on available indigenous social structures to ensure equity in relief provision during disasters.

Despite the HFA global initiative and a series of major disasters, including the 2010 flooding in Pakistan, the substantial reduction of disaster losses is still to be achieved (Oxley 2011). Effective integration of vulnerability assessment of the social, economic and ecological systems at various localities to the impacts of climate change and the strategies of adaptation and mitigation measures into sustainable development policies, planning and programming at all levels will ensure better outcomes to any future disaster.

Notes

1 Barrage built in 1932 for irrigation on River Indus near the city of Sukkur in the Sindh province
2 Glacier-made lake
3 A small island in river formed by the deposition of sediment in the water.
4 Dams can increase flooding upstream because of the backwater pressure as the upstream flow is impeded by structure.
5 Plantation of fast-growing tree species in line on field boundaries and sides of public roads, rivers and streams.
6 Community-based organisations.
7 National Disaster Management Authority.

References

ADB/WB – Asian Development Bank and World Bank. (2010). Pakistan Floods 2010: Preliminary Damage and Need Assessment. http://reliefweb.int/sites/reliefweb.int/files/resources/64AE3DC5BEDA4E18492577DA001FBE55-Full_Report.pdf (Accessed: 25 July 2014).Ahmed, R. (2011). *Remote Sensing for Delineation of Breaches along the Embankments of Indus River During Flood 2010*. Pakistan Space and Upper Atmosphere Research Commission, Karachi, Pakistan.

Ahmad, S. (2000). *Indigenous Water Harvesting Systems in Pakistan*. Water Resources Research Institute, National Agricultural Research Center, Islamabad.

Ahmad, Z. (2011). Impact of Alluvial Deposits on Soil Fertility During the Floods of 2010 in Punjab, Pakistan. Research Findings No. 26, International Potash Institute.

Anjum, S.A., Saleem, M.F., Cheema, M.A., Bilal, M.F. and Khaliq, T. (2012). An assessment to vulnerability, extent, characteristics and severity of drought hazard in Pakistan. *Pakistan Journal of Science*, 64(2), 85–96.

Archer, D.R. and Fowler, H.J. (2004). Spatial and temporal variation in precipitation in the Upper Indus Basin, global teleconnections and hydrological implications. *Hydrology and Earth System Sciences*, 8(1), 47–61.

Asif, I., Clift, P.D., Giosan, L., Tabrez, A.R., Tahir, M., Rabbani, M.M. and Danish, M. (2007). The geographic, geological and oceanographic setting of the Indus River. In Gupta, A. (ed.), *Large Rivers: Geomorphology and Management*. John Wiley & Sons, New York. 1.

Atta-ur-Rahman and Khan, A. N. (2013). Analysis of 2010-flood causes, nature and magnitude in the Khyber Pakhtunkhwa, Pakistan. *Natural Hazards*, 66(2), 887–904.

Atta-ur-Rehman, Khan, F., Moench, M., Malik, S., Sabbag, L. and MacClune, K. (2013). *Desk Study: Indus Floods Research Project*. Institute for Social and Environmental Transition, Boulder, CO.

Awan, S.A. (2003). The climate and flood-risk potential of northern areas of Pakistan. In *Water Resources in the South: Present Scenario and Future Prospects*. Islamabad, Pakistan: Commission on Science and Technology for Sustainable Development in the South – COMSATS. 87–100.

Azam, J.P. and Rinaud, J.D. (2000). Encroached Entitlements: Corruption and Appropriation of Irrigation Water in Southern Punjab (Pakistan). *Working paper 144*, SSRN.

B4A – Bytes for All, Pakistan. (2010). Watan at the Crossroads! Report on Watan Cards Programme in Pakistan. http://bytesforall.pk/cms/sites/default/files/Watan%20Card%20 Pakistan.pdf (Accessed: 14 September 2014).

Bisht, M. (2013). *Water Sector in Pakistan: Policy, Politics and Management*. Institute for Defense Studies and Analyses, New Delhi. 1.

Brooker, J. (2011). *Pakistan Floods: One Year on*. United Nations, Pakistan.

Budhani, A. and Gazder, H. (2011). *Land Rights and the Indus Flood, 2010–2011*. Rapid Assessment and Policy Review. Oxfam Research, Oxford.

Chaudhary, Q., Mahmood, A., Rasul, G. and Afzaal, M. (2009). Climate Change Indicators of Pakistan. Pakistan Meteorological Department Technical Report No. PMD-22/2009.

Cohen, R. (2014). Functions of Riparian Areas for Flood Control. http://www.mass.gov/eea/docs/dfg/der/riverways/riparian-factsheet-1.pdf (Accessed: 26 July 2014).

Collins, D. (1996). *Sediment Transport from Glacierized Basins in the Karakoram Mountains: Erosion and Sediment Yield: Global and Regional Perspectives*. International Association of Hydrological Sciences, Walligford, UK. 236.

Cornwell, K. and Hamidullah, S. (1992). Geomorphic evidence of catastrophic flooding along the middle Indus valley. *Geological Bulletin University of Peshawar*, 25, 113–121.

Elalem, S. and Pal, I. (2015). Mapping the vulnerability hotspots over Hindu-Kush Himalaya region to flooding disasters. *Weather and Climate Extremes*, 8, 46–58.

Fetherston, K., Naiman, R., and Bilby, R. 1995. Large woody debris, physical process, and riparian forest development in montane river networks of the Pacific Northwest. *Geomorphology*, 13, 133–144.

FFC – Federal Flood Commission. (2012). *Annual Flood Report, 2010*. Government of Pakistan.

Galetzka, J., Melgar, D., Genrich, J., Geng, J., Owen, S., Lindsey, E.O., . . . Maharjan, N. (2015). Slip pulse and resonance of the Kathmandu basin during the 2015 Gorkha earthquake, Nepal. *Science*, 4,349(6252), 1091–1095.

Gaurav, K., Sinha, R. and Panda, P.K. (2011). The Indus flood of 2010 in Pakistan: A perspective analysis using remote sensing data. *Natural Hazards*, 59, 1815–1826.

Giampaoli, P. and Aggarwal, S. (2010). Land Tenure and Property Rights in Pakistan: Failure to Address Land Tenure and Property Rights Grievances May Foster Support for the Taliban. USAID Issue Brief 4.

Gronewold, N. (2010). Is the flooding in Pakistan a climate change disaster? Devastating flooding in Pakistan may foreshadow extreme weather to come as a result of global warming. *Scientific American*. https://www.scientificamerican.com/article/is-the-flooding-in-pakist/#

Hanif, M. (2003). Management of water resources in South Asia. In *Water Resources in the South: Present Scenario and Future Prospects*. COMSATS Science and Technology, Islamabad, Pakistan. 3–20.

Hashmi, H.N., Siddiqui, Q.T.M., Ghumman, A.R., Kamal, M.A. and Mughal, H.R. (2012). A critical analysis of 2010 floods in Pakistan. *African Journal of Agricultural Research*, 7(7), 1054–1067.

Hewitt, K. (2005). The Karakoram anomaly? Glacier expansion and the 'elevation effect,' Karakorum Himalaya. *Mountain Research and Development*, 25(4), 332–340.

Hiroshi, P.S., Hasegawa, H., Fujiwara, S., Tobita, M., Koarai, M., Une, H. and Iwahashi, J. (2007). Interpretation of landslide distribution triggered by the 2005 Northern Pakistan earthquake using SPOT 5 imagery. *Landslides*, 4(2), 113–122.

Houze, R.A., Rasmussen, K.L., Medina, S., Brodzik, S.R. and Romatschke, U. (2011). Anomalous atmospheric events leading to the summer 2010 floods in Pakistan. *Bulletin of the American Meteorological Society*, 92(3), 291–298.

Hussain, S.S. (2010). *Pakistan 2010 Floods: Causes and Lessons Learnt*. Oxfam, Pakistan.

Imam, M. and Lohani, A. (2012). Beyond Water Conflict in the Indus Basin: Building Interprovincial Trust. Master thesis, Harvard University, Cambridge, MA.

Inam, A., Clift, P.D., Giosan, L., Tabrez, A.R., Tahir, M., Rabbani, M.M. and Danish, M. (2007). The geographic, geological and oceanographic setting of the Indus River. In Gupta, Avijit (ed.), *Large Rivers: Geomorphology and Management*. Wiley, New York. 333–345.

Iqtidar, H. (2010). *Why Do Natural Disasters Impact Some Areas More Than Others? Structural Improvement and the Floods in Pakistan*. Published on 15 September 2010 by the Social Science Research Council, Pakistan. http://itemsandissues.ssrc.org/pakistan/all/1 (Accessed: 26 July 2014).

Iturrizaga, L. (2008). Para-glacial landform assemblages in the Hindu Kush and Karakoram Mountains. *Geomorphology*, 95(1), 27–47.

Izhar-ul-Haq, D. (2014). Sediment Management of Tarbella Reservoir. Paper No. 733. Proceedings of the 72nd Annual Session 2011–2013, Pakistan Engineering Congress, Lahore, Pakistan.

Jodha, N. (2005). Economic globalisation and its repercussions for fragile mountains and communities in the Himalayas. *Global Change and Mountain Regions*, 23, 583–591.

Joshi, D.D. and Khan, A.A. (2009). Seismic vulnerability vis-à-vis active faults in the Himalaya – Hindukush Belt. *Journal of South Asia Disaster Studies*, 2(1), 197–243.

Khan, F.K. (2006). *Pakistan: Geography, Economy and People*. Oxford University Press, Lahore Pakistan.

Khan, A., Ayaz, M., Said, A., Ali, Z., Khan, H., Ahmad, N.N. and Garstang, R. (2010). *Rapid Assessment of Flood Impact on the Environment in Selected Affected Areas of Pakistan.* Pakistan Wetlands Programme and UNDP, Pakistan.

Khan, G.S. and Choudhry, A.K. (2007). Effect of spacing and plant density on the growth of Poplar (*Populus deltoides*) trees under agro-forestry system. *Pakistan Journal of Agricultural Science*, 44(2), 321–327.

Khan, S. (2011). *IWRM for Flood Management at the River Basin Level: Strategic Strengthening of Flood Warning and Management Capacity of Pakistan.* Islamabad, Pakistan: UNESCO, International Hydrological Programme.

Khan, S., Haq, M., Umar, M., Alam, N., Khan, M. U. and Masud, S. (2012). 2010 Floods in Pakistan: Damage assessment along Kabul River. *Journal of Himalayan Earth Science*, 45(2), 86.

Khan, T. (2010). Forest Management Paradigms and Resource Rights in Historical Perspective: Evidence from the Swat District, Pakistan. *Working paper 116*, Sustainable Development Policy Institute, Islamabad.

Khandekar, M. (2010). 2010 Pakistan floods: Climate change or natural variability? *Canadian Meteorological and Oceanographic Society Bulletin*, 38(5), 165–167.

Kirsch, T.D., Wadhwani, C., Sauer, L., Doocy, S. and Catlett, C. (2012). Impact of the 2010 Pakistan floods on rural and urban populations at six months. *PLoS Currents*, 4, e4fdfb212d2432.

Kreutzmann, H. (2006). Water towers for Pakistan: Irrigation practices entangled with ecology, economy and politics. *Geographische Rundschau*, 2(4), 48–55.

Lau, W.K. and Kim, K.M. (2012). The 2010 Pakistan flood and Russian heat wave: Teleconnection of hydro-meteorological extremes. *Journal of Hydrometeorology*, 13, 392–403.

Lewis, J. (2010). Corruption: The hidden perpetrator of under-development and vulnerability to natural hazards and disasters. *JAMBA: Journal of Disaster Risk Studies*, 3(2), 464–475.

Looney, R. (2012). Economic impacts of the floods in Pakistan. *Contemporary South Asia*, 20(2), 225–241.

Magsi, H. and Atif, S. (2012). Water management, impacts and conflicts: Case of Indus water distribution in Sindh, Pakistan. *International Journal of Rural Studies*, 9(2), 1–6.

Margolin, M. (1986). Propagating willow trees for soil erosion control. *Mother Earth News*. http://www.motherearthnews.com/nature-and-environment/trees-for-soil-erosion-zmaz86mazgoe.aspx#axzz3AIy8kptU (Accessed: 16 July 2014).

McGuire, C. (2012). *Why the 2010 Indus Floods Hit Pakistan so Hard.* International Development Research Center, Canada.

Meehl, G.A., Washington, W.M., Collins, W.D., Arblaster, J.M., Hu, A., Buja, L.E., . . . Teng, H. (2005). How much more global warming and sea level rise? *Science*, 307(5716), 1769–1772.

Mehari, A., van Steenbergen, F. and Schultz, B. (2007). Water rights and rules, and management in spate irrigation systems in Eritrea, Yemen and Pakistan. In Koppen, B. C. van and Butterworth, J. (eds) *Community-Based Water Law and Water Resource Management Reform in Developing Countries*, CABI, Walligford, UK and Cambridge, MA. 114–129.

Memon, N. (2011). *Root Causes of Floods. Dawn,* June 20, 2011.

Memon, N. (2014). *Malevolent Floods of Pakistan, 2010–2012.* Strengthening Participatory Organizations, Islamabad, Pakistan.

Mirza, B.B., Sharif, Z.M., Szombathova, N. and Zaujec, A. (2005). Rehabilitation of problem soils through environmental friendly technologies: Role of Sesbania and Phosphorus. *Agricultura Tropica Et Subtropica*, 38(1), 17–21.

Mohtadullah, K. C., Ata-ur-Rehman and Munir, C. M. (1992). Water for the 21st Century. No. 3. A Pakistan National Conservation Strategy, 3, IUCN.
Mustafa, D. and Wrathall, D. (2011). Indus basin floods of 2010: Souring of a Faustian bargain? *Water Alternatives*, 4(1), 72–85.
MWC. (2007). *Flood Management and Drainage Strategy*. Melbourne Water Corporation, Melbourne, Australia.
Naiman, R. J. and Décamps, H. (1997). The ecology of interfaces: Riparian zones. *Annual Review of Ecological Systems*, 28, 621–658.
Naseer, E. (2013). Pakistan's Water Crises. Spearhead Research Special Report, Pakistan.
Nation. (2010). 54 villages swept away in Upper Swat, Shangla floods: GCO. *The Nation*, August 10, 2010. http://nation.com.pk/politics/02-Aug-2010/54-villages-swept-away-in-Upper-Swat-Shangla-floods-GCO.
National Disaster Management Authority. (2011a). Pakistan: National Progress Report on the Implementation of the Hyogo Framework for Action (2009–2011). NDMA, Prevention Web.
National Disaster Management Authority. (2011b). A Review of the Pakistan Flood Relief and Early Recovery Response Plan (PFRERRP) Up to December 31, 2010. NDMA, 1–20, Prevention Web.
Nilsson. C. and Berggren, K. (2000). Alterations of riparian ecosystems caused by river regulation. *BioScience*, 50(9), 783–792.
Noor-ul-Haq. (2010). *Pakistan's Water Concerns*. Islamabad Policy Research Institute, Pakistan.
Oxley, M. (2011). Field note from Pakistan floods: Preventing future flood disasters. *JÀMBÁ: Journal of Disaster Risk Studies*, 3(2), 353–461.
PACF – Peoples Accountability Commission on Floods. (2012). Delayed cash compensation support jeopardizes the situation further for flood affected peoples 2010 and 2011 in Sindh. *Civil Society Floods Situation Report*, 16, 1–12.
Provincial Disaster Management Authority. (2012). *Contingency Plan Monsoon, 2012*. Provincial Disaster Management Authority, Khyber Pakhtunkhwa, Pakistan.
Provincial Disaster Management Authority. (2014). Daily Weather Report, 11 August 2014. Provincial Disaster Management Authority Khyber Pakhtunkhwa. Monsoon weather Rivers situation. (PDMA).
Rahim, I. (2010). Floods in mountain areas. *SGMOIK/SSMOCI Bulletin*, 31, 18–21.
Rahim, I. and Viaro, A. (2002). *Swat: An Afghan Society in Pakistan – Urbanization and Trends in a Tribal Environment*. Institute of Development Studies, Geneva, Switzerland.
Rasul, G., Chaudhry, Q. Z., Mahmood, A., Hyder, K. W. and Dahe, Q. (2011). Glaciers and glacial lakes under changing climate in Pakistan. *Pakistan Journal of Meteorology*, 8(15), 1–8.
Rasul, G., Mahmood, A., Sadiq, A. and Khan, S. I. (2012). Vulnerability of the Indus delta to climate change in Pakistan. *Pakistan Journal of Meteorology*, 8(16), 89–107.
Riaz-ul-Haq. (1994). Floods in Indus basin: Their control and management. *Paper 181, Symposium of Pakistan Engineering Congress*, 23, 9–40.
Schneider, J. F., Gruber, F. E. and Mergili, M. (2011). Recent Cases and Geomorphic Evidence of Landslide-Dammed Lakes and Related Hazards in the Mountains of Central Asia. *Proceedings of the second world landslide forum*. Rome. 1–6.
Shroder Jr, J., ed. (2004). *Himalaya to the Sea: Geology, Geomorphology and the Quaternary*. Routledge, New York.
Siddiqui, I. H. (2010). Re-Examining Flood Management Measures for Sindh. *Proceedings of the international workshop on floods in Pakistan-2010*, Lahore, Pakistan. 277 121–142.

Steimann B. (2003). Decentralisation & participation in the forestry sector of NWFP, Pakistan – The role of the state. Master thesis at the Department of Geography, University of Zurich.

Syvitski, J. and Brakenridge, G. (2013). Causation and avoidance of catastrophic flooding along the Indus River, Pakistan. *Geographical Society of America*, 23(1), 4–10.

Tariq, M.A. and van de Giesen, N. (2012). Floods and flood management in Pakistan. *Physics and Chemistry of the Earth, Parts A/B/C*, 47–48, 11–20.

United States Agency for International Development. (2010). *Property Rights and Resource Governance in Pakistan*. USAID, Washington, DC.

Wang, S.Y., Davies, R.E., Gillies, R. and Jin, J. (2011). Changing monsoon extremes and dynamics: Example in Pakistan. *NOAA Science and Technology Infusion Climate Bulletin*, National Weather Service, US. 61–68.

Wang, S.Y., Davies, R.E., Huang, W.R. and Gillies, R.R. (2011a). Pakistan's two stage monsoon and links with the recent climate change. *Journal of Geophysical Research: Atmospheres*, 116, D16.

Webster, P.J., Toma, V.E. and Kim, H.M. (2011). Were the 2010 Pakistan floods predictable? *Geophysical Research Letters*, 38, L04806.

White, S. (2011). *The 2010 Flooding Disaster in Pakistan: An Opportunity for Governance Reform or Another Layer of Dysfunction?* Center for Strategic and International Studies, Washington.

WHO – World Health Organization. (2010). *Pakistan Health Cluster – Floods in Pakistan*. World Health Organization, Cairo.

World Food Program. (2010). *Flood Impact Assessment*. World Food Program, Pakistan.

World Meteorological Organization. (2012). *WMO Statement on the Status of the Global Climate in 2011*. World Meteorological Organization, Geneva, Switzerland. WMO-No. 1085, 19.

4 Climate hazards and health in Asia

Ilan Kelman and Tim Colbourn

The largest and most populous continent, Asia, presents prominent challenges with respect to climate and health. While death tolls are notoriously unreliable and exact figures can be difficult to provide, some of the highest-fatality river floods have occurred in China, killing millions (Cai et al. 2001), while some of the highest-fatality storm surges (killing hundreds of thousands) and tornadoes (killing hundreds) have been in Bangladesh (Chowdhury et al. 1993; Paul 1998). The Himalayas experience deadly glacial lake outburst floods (Jain et al. 2012). Many of the low-lying atolls threatened by climate change lie in Asia, including Kiribati, the Maldives, the Marshall Islands and Tuvalu (IPCC 2013–2014).

Given the past history, the present situation and the future challenges and opportunities, examining climate hazards and health for Asia is important for bringing focus to a continent with major difficulties in this realm – but also with plenty to offer and to teach the world about addressing the challenges and creating opportunities. The latter can never be overemphasised due to the importance of learning from and exchanging with everyone, rather than assuming that certain groups inevitably need assistance from other groups.

Communities facing the water-related health consequences of melting glaciers in the Andes can exchange knowledge with those in the Himalayas. The low-lying main cities of Barbados, the Bahamas and Bermuda can share with places in the Pacific ideas and strategies about the psychological impacts of migration. Indian and Australian farmers can be tackling similar environmental health challenges during drought. This chapter describes some of these health and climate topics for Asia, important for the continent's future sustainability. The IPCC (2013–2014, p. 5) defines climate as:

> the average weather, or more rigorously, as the statistical description in terms of the mean and variability of relevant quantities over a period of time ranging from months to thousands or millions of years. The classical period for averaging these variables is 30 years, as defined by the World Meteorological Organization. The relevant quantities are most often surface variables such

> as temperature, precipitation, and wind. Climate in a wider sense is the state, including a statistical description, of the climate system.

From this definition, climate hazards are phenomena, trends or processes related to the climate, hence related to the atmosphere, which have the potential for interacting with human vulnerability to lead to a disaster. Examples are cold temperatures, droughts, floods, fog, hot temperatures, lightning, rain, sleet, snow, storm surges, tornadoes, tropical cyclones and wind. Some climate hazards have links beyond climate and the atmosphere, such as avalanches, glacial surges, landslides, rockslides and wildfires. Processes which drive many of these hazards – at times worsening them but sometimes reducing their intensity, frequency or magnitude – include climate change, climate variabilities and climate cycles such as the El Niño-Southern Oscillation and the Pacific Decadal Oscillation.

WHO (World Health Organisation 1946) defines 'Health [as] a state of complete physical, mental and social well-being and not merely the absence of disease or infirmity.' One implication is that 'health' refers to individual human beings, and WHO (1946) does state that 'health is one of the fundamental rights of every human being.' The definition's ethos, though, applies to communities as well as to the environment. While recognising that no community is homogenous (Walmsley 2006), community health indicators exist, such as in Sri Lanka (Jayasekara and Schultz 2007), that could be applied elsewhere in Asia, after contextualising them.

Additionally, WHO (2015, online) states that

> Environmental health addresses all the physical, chemical, and biological factors external to a person, and all the related factors impacting behaviours. It encompasses the assessment and control of those environmental factors that can potentially affect health. It is targeted towards preventing disease and creating health-supportive environments.

The impetus is on a healthy environment for a healthy society and vice versa, but the implication is ecosystem health within the context of ecosystems being part of a wider environment that helps to prevent disease and to support human and community health.

For examining climate and health, this book focuses on Asia. Examples range from the Middle East and eastern Turkey through the Arabian Peninsula and the Hindu Kush-Himalayan Region to the Far East as well as from the Russian Arctic to the islands of Asian waters. The latter include Singapore, Indonesia, the Philippines, Japan and small islands states such as the Marshall Islands, Nauru, Timor-Leste and Tuvalu. The next section describes climate change and its impacts on climate hazards, after which comes a section explaining how vulnerability causes health outcomes more than these hazards. Further case studies are provided in the following section to indicate the hazard-vulnerability balance. Then, the conclusion provides a proposed agenda for research and action.

Climate change affecting hazards in Asia

Defining and scoping climate change

Contemporary climate change has two official definitions, one from science and one from policy. The main international scientific body responsible for assessing and synthesising climate change science is the Intergovernmental Panel on Climate Change (IPCC). IPCC (2013–2014, p. 5) states:

> Climate change refers to a change in the state of the climate that can be identified (e.g., by using statistical tests) by changes in the mean and/or the variability of its properties, and that persists for an extended period, typically decades or longer. Climate change may be due to natural internal processes or external forcings such as modulations of the solar cycles, volcanic eruptions, and persistent anthropogenic changes in the composition of the atmosphere or in land use.

On the policy side, the main UN process for addressing climate change is the United Nations Framework Convention on Climate Change (UNFCCC) which defines climate change to be 'a change of climate which is attributed directly or indirectly to human activity that alters the composition of the global atmosphere and which is in addition to natural climate variability observed over comparable time periods' (UNFCCC 1992, Article 1, Paragraph 2). In summary, the science examines all changes to the climate irrespective of origin while the UN's policy process and measures consider climate change from only human origins.

For this chapter, the scientific definition is adopted, focusing on climate change's effects on climate hazards. While evidence exists for climate change altering non-climate-related hazards, such as earthquakes and volcanoes (McGuire 2013), the findings have large uncertainties, so this chapter is confined to climate hazards. Overall, there is no doubt that climate change affects climate hazards, including in Asia. How those hazards are affected involves major complexities, meaning that it would be challenging to give any hierarchy or ranking regarding importance or potential effects. Much depends on the local context.

Zoonotic disease

Costello et al. (2009) summarise the impacts of climate change on rodent-borne and vector-borne diseases. As temperatures increase, vectors and parasites tend to breed and mature more quickly. More life cycles are permitted within a given timeframe along with an increased rate of biting, each of which supports the spread of vector-borne diseases. In addition to these time factors, vector density over a given area tends to increase with temperature increasing. Meanwhile, vectors are able to survive at higher altitudes and latitudes than before because the climate in the new locations matches the vectors' environments in the original locations. Populations living at higher altitudes and latitudes have often never before dealt

with these vectors or pathogens, so their immunity and knowledge of countermeasures is minimal.

Costello et al. (2009) particularly highlight mosquito-borne and tick-borne diseases such as malaria, encephalitis and dengue fever, but their arguments likely apply to many other diseases such as Lyme disease, leptospirosis and West Nile Fever. The authors, though, indicate some climate hazard factors inhibiting vector-borne and rodent-borne diseases due to climate change. Where extreme event frequency or intensity increases, or where the environmental hazards change in nature, vector eggs and larvae could be harmed. Heavy rains can wash away vectors at many life cycle stages. Salinisation of water due to sea-level rise and coastal inundation could inhibit vectors needing freshwater or could force them to move inland or to higher elevations. Consequently, the infectious disease-related health impacts of climate change will be highly localised, depending on how specific climate hazards are affected locally – as well as measures taken to deal with disease.

Climate-related hazards

Climate change influences the frequency, severity, location and characteristics of weather events such as floods, droughts, temperature extremes and tropical cyclones. The complexities of the interactions between climate change and specific hazards in specific locations sometimes make attribution and projections challenging – again, also because measures taken to deal with the hazards can affect hazard parameters more than climate change.

For tropical cyclones – including for Pacific and Indian Ocean storms – it seems likely that climate change will lead to reduced frequency of formation but increased intensity once a tropical cyclone forms, with large uncertainties in the projections (Knutson et al. 2010). There are feedbacks amongst sea surface temperatures, stratospheric winds and the lessening temperature differential between the poles and the tropics (the poles warm faster), in addition to other factors, making analysis challenging. Empirical evidence from around the Pacific collected so far from the Federated States of Micronesia, Kiribati, Tuvalu and Vanuatu (Webb and Kench 2010; Rankey 2011; Ballu et al. 2012; Biribo and Woodroffe 2013) suggests that it is not inevitable that the low-lying islands will disappear due to sea-level rise. Instead, island responses will vary depending on many factors. In some places, such as Takuu and Papua New Guinea, sea-level rise is already causing inundation and salination problems leading to plans for resettlement.

Precipitation is expected to become much more intense in many locations around the world, including across Asia, due to climate change. Warmer air can hold more moisture, meaning that precipitation quantity and intensity can increase. That seems likely to lead to more flooding overall around the world (IPCC 2013–2014) with regional and seasonal variations. Meanwhile, as sea ice around the Arctic diminishes due to climate change, storms can produce more wave energy, which will likely exacerbate coastal erosion along the Russian Arctic. Melting permafrost will also significantly change the environments of the higher latitudes.

Such complexities impact other hazards and their potential casualties. Increased precipitation under climate change would be expected to increase the frequency and magnitude of landslides, but it is not straightforward. For a landslide, rock-slide, mudslide or other mass movement to occur, there must be material to slide or move. An initial increase in frequency due to climate change has the possibility of using up much of a slope's slideable material, leading to smaller slides in the future because less material can build up before the slide occurs.

Vulnerability determining health outcomes

The discussion in the previous section covered the effects of climate change on climate hazards with indications of resultant casualties. This section delves deeper into the latter exploring the root causes and underlying reasons of why casualties occur due to climate hazards. Mortality projections under climate change are provided, followed by a description of vulnerability and related concepts. That sets the stage for exploring causal chains from hazard and vulnerability to mortality, using weather, food, farming and traditional knowledge to illustrate the health impacts.

WHO (2014) estimates global mortality resulting from climate change in 2030 and 2050 via effects from heat, coastal flooding, diarrhoeal disease, malaria, dengue and undernutrition, all impacts of which will be felt in Asia. That study assumes no major discontinuities in trends; that is, no major societal collapses, wars or famines, but still estimates 92,781 (with low estimate of 46,314 and a high estimate of 140,100) extra deaths per year from these causes in Asia in 2050, with South Asia hardest hit (WHO 2014). Aside from hazards not considered such as cold and slides, it is important to examine the assumed causal chain from climate change to climate hazards (discussed above) and then to casualties (discussed here).

Climate hazards alone do not cause injuries and deaths. Vulnerability – the social and political processes that put people and communities in harm's way without adequate resources or options to avoid harm – must be present individually or collectively (Hewitt 1983; Lewis 1999; Wisner et al. 2004, 2012; Gaillard 2010). Meanwhile, resilience processes can reduce casualties and other detrimental health effects by supporting people and communities in adjusting or adapting to climate hazards and changes in their regimes. *Resilience* is a highly contested term across many disciplines which has moved away from a basic ecological definition of being able to withstand shocks while maintaining equilibrium and much more towards fundamental tenets of disaster risk reduction in terms of being able to build and maintain safe and healthy communities (Manyena 2006; Lewis and Kelman 2010; Alexander 2013; Manyena and Gordon 2015).

Because so much about vulnerability and resilience is contextual, they are parallel rather than opposite processes, expressing different manifestations of fundamental social and political processes which permit society to deal with hazards – or not. Sometimes, the same action can breed both vulnerability and resilience – often to different sectors of the community. For instance, some groups in Pakistan and Afghanistan oppose girls attending school. School buildings can be made centres

for disaster risk reduction and be known in the community as a safe post-disaster location with emergency supplies (Ronan and Johnston 2006). Where there is gender disparity in school attendance, implementing 'safe schools' programmes could increase boys' resilience, while increasing girls' vulnerability relative to the boys, all to the same flood. Often, the degree in which vulnerability and resilience are present to a specific environmental phenomenon for a specific community group can contribute to determining how hazardous that environmental phenomenon becomes.

Considering river flooding in Bangladesh (Webster et al. 2010) or a heat wave in Seoul such as in 1994 (Kyselý and Kim 2009), these hazards have long been a feature of life in these places. Some people have been able to cope with them to a certain degree, reducing their vulnerability and increasing their resilience, such as the people living on river islands in Bangladesh where flooding is a way of life (Sarker et al. 2003). Others run into major difficulties, such as those aged over 70 or under 14 in South Korea who were not supported to deal with the heat (Kyselý and Kim 2009). In Bangladesh, if floods increase in frequency, then the people used to seasonal floods might not be able to deal with the changes without external support – or, as Sarker et al. (2003) imply, increased erosion would be problematic. In Seoul, if the emergency management system and social networks had responded in 1994, identifying those who were vulnerable and bringing them to relief centres or providing them with fans and support to pay the electricity bill (amongst other measures), then the death toll would not have been so high. Consequently, even in cases where changes to extreme weather can be directly linked to climate change, the causal pathway to the weather casualties is not clear.

Causal chains are also not straightforward for climate change leading to health impacts due to changes in food/water quality and availability, or due to psychological effects. As average sea level rises, mainly due to the increasing ocean temperature but with contributions from melting ice, water supplies are becoming salinised, also through coastal inundation. Freshwater patterns are changing as well, especially as precipitation and storms change (IPCC 2013–2014), so the link to water-borne disease can be hard to determine. Regarding food, depending on localised changes, it might be harder or easier to harvest from the land or ocean. Due to a preference for imported food, many Pacific islands currently experience an obesity epidemic leading to non-communicable diseases such as diabetes (Asia Pacific Cohort Studies Collaboration 2007). If transport costs increase under climate change, then one possible pathway is less dependency on imports, which leads to improved health conditions, although with the potential for undernutrition if local food resources prove to be inadequate. The trade-offs and balances are complex with the influence of climate change and climate hazards potentially being small compared to vulnerability of and choices by individuals and communities.

Climate hazards have always been part of farming, including crop and livestock mortality. For example, a review of cotton crop failure linked to farmer suicides in India demonstrated subtle and complex relationships amongst multiple factors (Gruère and Sengupta 2011) with further concerns about why Indian farmers selected – or were encouraged to select – specific cotton crops rather than

focusing on food crops or a balance between subsistence and cash crops. That is important since traditional knowledge systems across Asia have developed over time, and have shown to be adaptable, supporting food security irrespective of climate hazards. Examples are systems for sharing water resources and coordinating agricultural activities, landscape management and crop diversity (Moles 1989; Mohamma 1992).

Those traditional systems did not always prevent disasters, yet they provided a baseline for dealing with environmental changes, variabilities and trends, permitting flexibility and adaptability – hallmarks of reducing vulnerability and increasing resilience. With climate under climate change projected to move beyond humanity's collective memory, traditional knowledge is becoming less relevant for trying to deal with climate hazards, especially given the major social changes which Asia is also experiencing. Impacts on livelihoods from climate change and climate hazards are inextricably linked with impacts from the undermining of traditional knowledge, changing values, environmental health affected by infrastructure development and pollution, lack of societal interest in supporting people for disaster risk reduction including climate change adaptation, poor social services to support people at risk of suicide and numerous other vulnerability factors – alongside factors which support resilience such as good governance and education opportunities.

Climate has wide-ranging impacts on health, positive and negative. Decoupling those from other processes – such as choices by those with resources to support or not to support adjustment or adaptation to changing conditions – is not always unambiguous, with chronic vulnerabilities often dominating direct impacts of climate hazards.

Further case studies

This section provides further examples of health impacts of climate hazards in Asia, categorised by hazard (water, temperature and specific diseases) and by consequence (undernutrition and mental health). The discussion cannot be comprehensive on each topic, merely providing some aspects to consider and aiming to meld that with the climate change and vulnerability discussion from the previous two sections.

Water: storms, precipitation, floods and droughts

Flooding, precipitation and storms lead to casualties through impacts as diverse as drowning; physical trauma from debris; electrocution; increases in faecal-oral, vector-borne and rodent-borne disease and loss of livelihoods and shelter (Ahern et al. 2005). Bangladesh is an example of an Asian location affected by flooding, with 70 per cent of the population living in flood-prone areas, and 26 per cent at risk from cyclones, although interventions, including improved disaster risk reduction and management, storm shelters and warning systems, have demonstrated how much mortality can be reduced (Cash et al. 2013).

Overall, data from South Asian floods show that more women than men tend to die in most types of floods, mainly due to traditions which increase women's vulnerability (for example, Bern et al. 1993; Chowdhury et al. 1993). Women are often not permitted to learn how to swim, might not be permitted to be outside alone and tend to wear clothes which make it hard to survive in water (taking off their clothes in order to survive would not usually be considered as an option). Women also tend not to get enough to eat, making them undernourished and weaker. Other factors have not been fully studied but are understood to exist, such as the potential fear of sexual violence in evacuation locations or menstruating women not being permitted to be in public areas.

In addition to deaths, injuries and mental health issues, water-related hazards can affect environmental health by exacerbating endemic disease. Examples from flooding and rainfall were outbreaks of leptospirosis in the Philippines in 2009 (Amilasan et al. 2012) and human fascioliasis in Iran in 1988 and 1999 (Salahi-Moghaddam et al. 2011). Health impacts of drought were seen in Iran in 1999–2000 with increases in human skin and eye infections, more animal parasites and livestock deaths (Salami et al. 2009).

Sea-level rise is expected to affect low-lying coastlines, potentially forcing people to move: Bangladesh and the Pacific islands are usually the locations most likely to be affected the soonest (IPCC 2013–2014). Managed evacuation and resettlement can avoid many physical health impacts, but mental health consequences have not been fully investigated, particularly given the extent to which many people are connected to their home and land. In the Pacific, one's identity is closely tied up with one's land and ancestral burial grounds (Ward and Kingdon 2007). Abandoning that and moving elsewhere, never to return, would not be simple for everyone, even if resettlement were fully resourced. Simultaneously, environmental health is being affected through increased salinisation and changes to ecosystems as new species arrive and some endemic species struggle in the new climate regime.

More widely, water-related hazards affect environmental health across Asia. Too much water, too little water, high winds, salt water and contaminated water affect the natural and built environments and their intersections. For instance, mould can build up inside properties, disease vectors can be promoted or inhibited and ecosystems are rapidly changed. Specific examples range from road dust being transported by sandstorms to Beijing, thereby increasing air pollution (Han et al. 2007), through to drinking water source contamination after a cyclone struck Bangladesh (Hoque et al. 1993).

Temperature: heat and cold

Very old and very young people are most affected by heat-related mortality but outdoor exposure, whether working in fields or walking kilometres to fetch water, increases impacts. In Japan, the relative risk of mortality in over-65s was estimated to rise to 1.5 as daily maximum temperatures exceeded optimal temperatures by 20°C (Honda et al. 2014). A model for Seoul, Beijing, Tokyo and Taipei estimated

links between air temperature and mortality from respiratory and cardiovascular diseases (Chung et al. 2009).

Mortality and morbidity from heatwaves are likely to increase under climate change (Kravchenko et al. 2013). Health impacts from heat stress will depend on physiological adaptation to higher average temperatures as well as living and working conditions. Those with abodes and/or workplaces that do not allow sufficient protection from heat stress – typically the poorest people – will be more affected, as will other vulnerable people including children, the elderly and people with chronic diseases (Kravchenko et al. 2013).

Productivity, especially related to outdoor work, such as agricultural labour, which is the mainstay for many rural subsistence farmers throughout Asia, will be affected by rising average temperatures and heat extremes. This, in turn, could have knock-on effects for nutrition and other health indicators via impacting local livelihoods for food and water. An empirical study of heat stress in Indian rice farmers estimated a 5 per cent drop in productivity per 1°C increase in heat exposure (Sahu et al. 2013).

Cold extremes have health impacts, such as people dying from lack of warmth as well as influenza being promoted in cold temperatures. In 2007–2008, temperatures in Dushanbe, Tajikistan remained below freezing for over a month, taxing the country's poorly maintained electricity infrastructure, impacting livelihoods as electricity-dependent businesses and factories shut down, promoting disease due to the cold and people living in a single room to keep warm and leading to other health impacts such as poor sanitation and carbon monoxide poisoning (Kelly 2012). In all these factors, the climate hazard of cold exacerbated and exposed pre-existing vulnerability conditions rather than being the cause of mortality and morbidity.

Diarrhoeal disease

Although diseases not present in an area before a disaster rarely emerge afterwards, the risk of diarrhoeal disease and other infectious diseases can increase in endemic areas. Reasons include pit latrine overflow, lack of clean water to drink and with which to wash, poor sanitation facilities and lack of personal hygiene after evacuating, overcrowding so that people are in close proximity to one another, insufficient food and cooking apparatus and lack of access to health services (which may have collapsed) including vaccination (Kouadio et al. 2012).

Diarrhoeal disease is projected to increase in Asia in the short term with rising temperatures under climate change, by an estimated 3–10 per cent per 1°C temperature rise (WHO 2014). Nonetheless, even under low socio-economic growth scenarios and with the temperature increases due to climate change, diarrhoeal disease in Asia is projected to be lower in 2050 than in 2030, primarily due to assumed improvements in water, sanitation and public health provision (WHO 2014). These projections again assume no major discontinuities in trends, but illustrate that complicated interactions amongst many factors can sometimes lessen the consequences climate hazards could have.

Malaria and dengue

Flooding and heavy rains – via the stagnant water mosquito breeding grounds left behind and population displacement from non-endemic to endemic areas – and increasing temperatures (see Costello et al. 2009) can increase risk of malaria and dengue fever in Asia (Kouadio et al. 2012; WHO 2014), balanced with the factors noted earlier which wash away larvae or inhibit mosquito breeding. In fact, sometimes mosquitoes are not a problem the first few days after a tropical cyclone, which leaves some time to implement prevention measures before they return.

Malaria transmission requires ambient temperatures of at least 16°C and appears to be optimal at 28–30°C, similar to the transmission of dengue; while mosquito survival declines above 35°C and is not usually possible at 40°C (WHO 2014). Nonetheless, increasing temperatures could result in adaptation and survival of the vector and parasite at higher temperatures (Caminade et al. 2014). Historically, malaria transmission and consequent morbidity and mortality have decreased as socio-economic conditions improve; indeed, malaria-type diseases were eradicated in Europe and North America. Nevertheless, despite expected improvements in socio-economic conditions across Asia, WHO (2014) estimates that the increased average ambient temperatures resulting from climate change will lead to more than an additional 200 million people being at risk of malaria in South and Southeast Asia in coming decades, with an even higher number at risk due to population increases.

Dengue fever is also affected through increased risk due to population growth and decreased risk due to socio-economic conditions improving in Asia, including piped water reducing use of the water storage containers in which dengue-carrying *A. aegypti* mosquitoes typically breed (WHO 2014). Increases in temperature and precipitation due to climate change in Asia are still expected to increase the number of people at risk of dengue fever overall (WHO 2014).

Undernutrition

Undernutrition is ultimately caused by a lack of calories, but has complex, underlying causes related to the production, availability and access to food. Hence, it is linked to poverty, trade, water and sanitation since diarrhoeal disease exacerbates conditions. Undernutrition is tackled, amongst other actions, through women's education and empowerment, such as knowledge of diets for children and ability to feed children properly (WHO 2014). Undernourished children are more vulnerable to infectious diseases and other health problems, including poor physical and cognitive development (Black et al. 2008).

Climate change is likely to adversely affect crop production, particularly in seasonally dry and low-latitude regions due to warming and a changing regime of climate hazards, while sea-level rise could salinise lands, making it necessary to change crops entirely (IPCC 2013–2014; WHO 2014). East Asia has already seen impacts on food production systems and food security including through rice

yields, vegetables, fruit and livestock. Attribution of observed crop, livestock, food and water trends directly to climate change is complex, especially given poorly understood responses of crops to increasing atmospheric carbon dioxide concentrations and the positive effects of adaptation measures (IPCC 2013–2014). Meanwhile, although any conclusions are tenuous, more northern latitudes in Asia might experience productivity increases due to longer growing seasons and warming temperatures.

Models (Lloyd et al. 2011; WHO 2014) estimate that, in South Asia in 2030, the effects of climate change on crop production, global food trade and consequent national calorie availability and distribution will yield an additional 1,100,000 children under the age of five who are moderately stunted and 900,000 who are severely stunted, with approximately 20,000 children dying from this malnutrition. South Asia has higher mortality relative to the number of cases of undernutrition due to additional vulnerabilities related to poverty and inadequate food distribution, indicating that climate hazards and the effects of climate change are only one factor amongst many.

Mental health impacts

There has been limited empirical research to date, especially in Asia, on the impact of climate change on mental health. A review by Bourque and Willox (2014) suggests that climate change could increase psychological distress, depression and anxiety as well as addiction and suicide rates, especially for those already vulnerable and those with pre-existing mental health conditions. Again, the key is vulnerability and pre-existing conditions, with climate hazards influencing existing problems but not necessarily causing completely new difficulties.

Environmental changes are already occurring rapidly in some areas such as the Arctic, namely northern Russia, and high altitudes, especially the Himalayas, with expected major mental health consequences for those dependent on natural resource–based livelihoods. A study from Bangladesh which was able to sample 162 children who had been assessed six months before a flood and five months afterwards found an increase from zero to almost 10 per cent in aggressive behaviour and also that 34 per cent of 134 children who did not wet their bed before, now did (Durkin et al. 1993). The authors consider these symptoms to be 'post-traumatic stress' and attribute them to the flood. Ahern et al. (2005) document increased anxiety, depression and post-traumatic stress disorder from floods from several other studies, although they found no evidence for an increase in suicides. Conversely, if rates of flooding increase, then inundation might become a regular part of life rather than being unusual and extreme, thereby decreasing adverse mental health impacts.

In addition to direct effects on mental health, the impact of climate hazards can exacerbate mental health problems by disrupting or stopping access to mental health services where they exist (Bourque and Willox 2014).

Agenda for research and action

From the above discussion, specific topics emerge as being prominent in a research and action agenda regarding climate hazards, health and Asia. Six main suggestions are provided here.

1 Climate change as a hazard driver, enhancer or diminisher, rather than as a cause of health impacts itself.

Climate change in itself does not cause health impacts, but human failure to respond to climate change and its consequences does. To develop and enact effective policies for Asia, it is important to move beyond the rhetoric of climate change as a direct cause and, instead, to understand better the causal sequence of how climate change alters environmental hazards, cycles, trends, processes and conditions, which in turn can expose already existing vulnerabilities or resilience, yielding positive and negative health impacts. Given the complexity of this causal chain and the large environmental differences in Asia – consider, for instance, megacity, desert, mountain, forest and atoll contexts – more research is needed to explore the causal chain for specific contexts and to build up a repertoire of comparative analyses from which general lessons and transferability/non-transferability options could be explored.

2 Causes and circumstances of individual morbidity and mortality.

Large gaps exist in understanding who is physically affected by climate hazards, how they are affected, and why their vulnerability is not addressed leading to death or injury in specific hazard instances. This is particularly the case in Asia where disaster deaths can be reported in the hundreds or tens of thousands with limited scope to investigate and explain each death on an individual basis, as has been completed for floods in Europe and the United States (for example, Jonkman and Kelman 2005) amongst others.

An additional challenge is that disaster mortality studies tend to focus on people whose death is clearly linked to a specific hazard, often looking at hazard as the only cause of death because such data are relatively unambiguous and comparatively easy to collect. Yet many deaths occur months or years after a specific hazard manifested or had multiple inputs into the death. The cause might be physical, for example someone in a coma who was not evacuated, but is often mental, due to stress, lack of post-disaster support or lack of financial or social resources to cope with difficulties. Because establishing cause and effect in such cases is challenging, these deaths frequently remain unnoticed and uncounted, becoming part of a disaster's hidden cost (see also point 4).

3 Improvements in mapping out and modelling cause and consequence pathways amongst hazards, vulnerabilities and health outcomes.

At the first order, hazard and vulnerability are generally considered to be independent variables, but that neglects the interactions amongst them. For example, people living in a hilly region in Bhutan experiencing a cold snap might not be able to afford increased electricity costs – or might not have electricity – so their vulnerability to cold forces them to cut down trees for burning. That amplifies the likelihood and intensity of flood and slide parameters, exposing them later to worsening climate hazards. Better understanding of the interactions and feedbacks can contribute to developing policy and practice which reduces vulnerabilities and improves health without exacerbating hazards.

4 Mental health impacts.

Studies of mental health impacts of climate hazards are few for Asia, yet this topic is essential for gaining a full picture of climate hazards and health. That relates to the psychological effects of specific climate hazards and responses to them, but can also be connected to climate change more directly. In places such as high-altitude communities and low-elevation coastal communities, the changing environment means that traditional knowledge is already becoming outdated to some degree, as seasons and biota shift. The psychological effects include the loss of an anchor connecting community with environment as well as, in places, a lack of understanding of why it is harder to provide food and livelihoods from the environment. If migration is required, then the loss of identity, culture and language, alongside the disorientation of moving from a place which might rarely have been left before, could be psychologically devastating. Such studies for Asia have rarely been conducted, but are starting to emerge for island and mountain communities, although they require a wider scope.

5 Environmental health impacts.

The interplay between environmental health and human health is understudied, from nomadic pastoralists in Mongolia dealing with shifting seasons through to pollution and proximity to waste management sites in urban settings. For Asia, little work connects or integrates environmental health indicators, community health indicators and individual health indicators. With the advent of ecosystem-based adaptation as a process for dealing with climate change, although noting that it is not different from ecosystem-based management, more work regarding environmental health impacts of climate change is expected and needed. Tackling this topic would also contribute to expanding the understanding and implementation of health interventions, by ensuring that these responses do not highlight only humans, but also factor in community surroundings including the built and natural environments.

6 Evaluate the reliability, validity and sustainability of new knowledge forms and their combinations.

Despite social and environmental changes across Asia, traditional knowledge, local knowledge, vernacular knowledge and indigenous knowledge have rich

contributions to make for improving health and for reducing vulnerability and building resilience in the face of climate hazards. These knowledge forms can be combined with external knowledge forms (see one framework in Mercer et al. 2010), such as science and professions providing remote sensed data, engineering analyses, historical interpretations and focus group data amongst many others. Analytical and conceptual models can be improved without relying exclusively on different forms of modelling, while new approaches for collecting quantitative and qualitative data can be developed, tested and applied.

Asia faces many challenges with respect to climate change and health. These challenges can become opportunities to reduce current vulnerabilities and to build resilience, including beyond climate; to improve the current state of human, community and environmental health; and to prepare Asia and the world for the changes to hazards expected under climate change. The key messages are the complexities of the interactions, the balance amongst positive and negative outcomes and the difficulties with assuming specific causal lines.

Those messages are illustrated by the main points of climate change influencing zoonotic and other infectious diseases along with climate-related hazards, although causality from climate change to health outcomes is not straightforward. Instead, vulnerabilities determine health outcomes with much being contextual, such as the efforts which have been put into building resilience, implementing disaster risk reduction processes including climate change adaptation, and supporting food, water and livelihood security. Decoupling climate change effects from all other processes is not easy, with undernutrition, mental health and environmental health demonstrating the interactions.

Providing relative magnitudes of importance or providing a hierarchy of concern for all these topics encounters credibility concerns for two reasons. First, data have often not yet been collected whilst projections have large uncertainties. Second, the importance usually depends on context. For instance, what applies to Phnom Penh might not apply to Siem Reap, Cambodia, Southeast Asia or the entire continent. These concerns are not a call to avoid general lessons and messages, as those have been provided here. Instead, it is important to accept that sometimes qualitative descriptions can provide the most robust ways forward.

Ultimately, climate change is only one factor (and frequently not the dominate one) amongst many regarding health, vulnerability and development. Where climate change ranks compared to other concerns is contextual, often with measures taken to deal with climate change and other concerns being more important than the projections for climate change impacts. Rather than highlighting climate change, by tackling vulnerability overall, it is possible for good health outcomes to emerge, irrespective of how the climate changes.

References

Ahern, M., Kovats, R. S., Wilkinson, P., Few, R. and Matthies, F. (2005). Global health impacts of floods: Epidemiologic evidence. *Epidemiologic Reviews*, 27, 36–46.

Alexander, D.E. (2013). Resilience and disaster risk reduction: An etymological journey. *Natural Hazards and Earth Systems Sciences*, 13, 2707–2716.
Amilasan, A.T., Ujiie, M., Suzuki, M., Salva, E., Belo, M.C.P., Koizumi, N., . . . Ariyoshi, K. (2012). Outbreak of leptospirosis after flood, the Philippines, (2009). *Emerging Infectious Diseases*, 18, 91–94.
Asia Pacific Cohort Studies Collaboration. (2007). The burden of overweight and obesity in the Asia–Pacific region. *Obesity Reviews*, 8(3), 191–196.
Ballu, V., Bouin, M.N., Siméoni, P., Crawford, W.C., Calmant, S., Boré, J.M., . . . Pelletier, B. (2012). Comparing the role of absolute sea-level rise and vertical tectonic motions in coastal flooding, Torres Islands (Vanuatu). *Proceedings of the National Academy of Sciences*, 108, 13019–13022.
Bern, C., Sniezek, J., Mathbor, G.M., Siddiqi, M.S., Ronsmans, C., Chowdhury, A.M., . . . Glass, R.I. (1993). Risk factors for mortality in the Bangladesh cyclone of 1991. *Bulletin of the World Health Organization*, 71(1), 73–78.
Biribo, N. and Woodroffe, C.D. (2013). Historical area and shoreline change of reef islands around Tarawa Atoll, Kiribati. *Sustainability Science*, 8, 345–362.
Black, R.E., Allen, L.H., Bhutta, Z.A., Caulfield, L.E., de Onis, M., Ezzati, M., . . . Rivera, J. (2008). Maternal and child undernutrition: Global and regional exposures and health consequences. *Lancet*, 371, 243–260.
Bourque, F. and Willox, A.C. (2014). Climate change: The next challenge for public mental health. *International Review of Psychiatry*, 26, 415–422.
Cai, S., Chan, N.W., Kung, H.-T. and Liu, P.S. (2001). Management of flood disasters in the Jianghan Plain, China. *Disaster Prevention and Management*, 10(5), 339–348.
Caminade, C., Kovats, S., Rocklov, J., Tompkins, A.M., Morse, A.P., Colón-González, F.J., . . . Lloyd, S.J. (2014). Impact of climate change on global malaria distribution. *Proceedings of the National Academy of Sciences of the United States of America*, 111, 3286–3291.
Cash, R.A., Halder, S.R., Husain, M., Islam, M.S., Mallick, F.H., May, M.A., . . . Rahman, M.A. (2013). Reducing the health effect of natural hazards in Bangladesh. *Lancet*, 382, 2094–2103.
Chowdhury, A.M.R., Bhuyia, A.U., Choudhury, A.Y. and Sen, R. (1993). The Bangladesh cyclone of 1991: Why so many people died. *Disasters*, 17(4), 291–304.
Chung, J.-Y., Honda, Y., Hong, Y.-C., Pan, X.-C., Guo, Y.-L. and Kim, H. (2009). Ambient temperature and mortality: An international study in four capital cities of East Asia. *Science of the Total Environment*, 408(2), 390–396.
Costello, A., Abbas, M., Allen, A., Ball, S., Bell, S., Bellamy, R. . . . Patterson, C. (2009). Managing the health effects of climate change. *Lancet*, 373, 1693–1733.
Durkin, M.S., Khan, N., Davidson, L.L., Zaman, S.S. and Stein, Z.A. (1993). The effects of a natural disaster on child behaviour: Evidence for post-traumatic stress. *American Journal of Public Health*, 83, 1549–1553.
Gaillard, J. (2010). Vulnerability, capacity, and resilience: Perspectives for climate and disaster risk reduction. *Journal of International Development*, 22(2), 218–232.
Gruère, G. and Sengupta, D. (2011). Cotton and farmer suicides in India: An evidence-based assessment. *Journal of Development Studies*, 47(2), 316–337.
Han, L., Zhuang, G., Chenga, S., Wang, Y. and Li, J. (2007). Characteristics of re-suspended road dust and its impact on the atmospheric environment in Beijing. *Atmospheric Environment*, 41(35), 7485–7499.
Hewitt, K., ed. (1983). *Interpretations of Calamity from the Viewpoint of Human Ecology*. Allen & Unwin, London.

Honda, Y., Kondo, M., McGregor, G., Kim, H., Guo, Y.-L., Hijioka, Y., . . . Kovats, R.S. (2014). Heat-related mortality risk model for climate change impact projection. *Environmental Health and Preventive Medicine*, 19, 56–63.

Hoque, B.A., Sack, R.B., Jahangir, A.M., Hazera, N., Siddiqi, M. and Nahid, A. (1993). Environmental health and the 1991 Bangladesh cyclone. *Disasters*, 17(2), 143–152.

IPCC (Intergovernmental Panel on Climate Change). (2014). *Fifth Assessment Report*. IPCC, Geneva.

Jain, S.K., Lohani, A.K., Singh, R.D., Chaudhary, A. and Thakural, L.N. (2012). Glacial lakes and glacial lake outburst flood in a Himalayan basin using remote sensing and GIS. *Natural Hazards*, 62(3), 887–899.

Jayasekara, R.S. and Schultz, T. (2007). Health status, trends, and issues in Sri Lanka. *Nursing and Health Sciences*, 9, 228–233.

Jonkman, S.N. and Kelman, I. (2005). An analysis of causes and circumstances of flood disaster deaths. *Disasters*, 29, 75–97.

Kelly, C. (2012). Cold extremes in Tajikistan. In Wisner, B., Gaillard, JC and Kelman, I. (eds.), *Handbook of Hazards and Disaster Risk Reduction*. Routledge, Abingdon, 276.

Knutson, T.R., McBride, J. L., Chan, J., Emanuel, K., Holland, G., Landsea, C., . . . Sugi, M. (2010). Tropical cyclones and climate change. *Nature Geoscience*, 3, 157–163.

Kouadio, I.K., Aljunid, S., Kamigaki, T., Hammad, K. and Oshitani, H. (2012). Infectious disease following natural disasters: Prevention and control measures. *Expert Review of Anti-Infective Therapy*, 10, 95–104.

Kravchenko, J., Abernethy, A.P., Fawzy, M. and Lyerly, H.K. (2013). Minimization of heat-wave morbidity and mortality. *American Journal of Preventive Medicine*, 44, 274–282.

Kyselý, J. and Kim, J. (2009). Mortality during heat waves in South Korea, 1991 to 2005: How exceptional was the 1994 heat wave? *Climate Research*, 38(2), 105–116.

Lewis, J. (1999). *Development in Disaster-Prone Places: Studies of Vulnerability*. Intermediate Technology Publications, London.

Lewis, J. and Kelman, I. (2010). Places, people and perpetuity: Community capacities in ecologies of catastrophe. *ACME: An International E-Journal for Critical Geographies*, 9, 191–220.

Lloyd, S.J., Kovats, R.S. and Chalabi, Z. (2011). Climate change, crop yields, and undernutrition: Development of a model to quantify the impact of climate scenarios on child undernutrition. *Environmental Health Perspectives*, 119, 1817–1823.

Manyena, S.B. (2006). The concept of resilience revisited. *Disasters*, 30, 433–450.

Manyena, S.B. and Gordon, S. (2015). Bridging the concepts of resilience, fragility and stabilisation. *Disaster Prevention and Management*, 24(1), 38–52.

McGuire, B. (2013). *Waking the Giant: How a Changing Climate Triggers Earthquakes, Tsunamis, and Volcanoes*. Oxford University Press, Oxford.

Mercer, J., Kelman, I., Taranis, L. and Suchet, S. (2010). Framework for integrating indigenous and scientific knowledge for disaster risk reduction. *Disasters*, 34(1), 214–239.

Mohammad, N., ed. (1992). *Dynamics of Agricultural Development*. Concept, New Delhi.

Moles, J.A. (1989). Agricultural sustainability and traditional agriculture: Learning from the past and its relevance to Sri Lanka. *Human Organization*, 48(1), 70–78.

Paul, B.K. (1998). Coping with the 1996 tornado in Tangail, Bangladesh: An analysis of field data. *The Professional Geographer*, 50(3), 287–301.

Rankey, E.C. (2011). Nature and stability of atoll island shorelines: Gilbert Island chain, Kiribati, equatorial Pacific. *Sedimentology*, 58, 1831–1859.

Ronan, K. and Johnston, D. (2006). *Promoting Community Resilience in Disasters: The Role for Schools, Youth, and Families*. Springer, New York.

Sahu, S., Sett, M. and Kjellstrom, T. (2013). Heat exposure, cardiovascular stress and work productivity in rice harvesters in India: Implications for a climate change future. *Industrial Health*, 51, 424–431.

Salahi-Moghaddam, A., Habibi-Nokhandam, M. and Fuentes, M.V. (2011). Low-altitude outbreaks of human fascioliasis related with summer rainfall in Gilan province, Iran. *Geospatial Health*, 6, 133–136.

Salami, H., Shahnooshi, N. and Thomson, K. J. (2009). The economic impacts of drought on the economy of Iran: An integration of linear programming and macroeconometric modelling approaches. *Ecological Economics*, 68(4), 1032–1039.

Sarker, M.H., Huque, I., Alam, M. and Koudstaal, R. (2003). Rivers, chars and char dwellers of Bangladesh. *International Journal of River Basin Management*, 1(1), 61–80.

UNFCCC (United Nations Framework Convention on Climate Change). (1992). *United Nations Framework Convention on Climate Change*. UNFCCC, Bonn.

Walmsley, J. (2006). The nature of community: Putting community in place. *Dialogue*, 25, 5–12.

Ward, R.G. and Kingdon, E. (2007). *Land, Custom and Practice in the South Pacific*. Cambridge University Press, Cambridge.

Webb, A.P. and Kench, P.S. (2010). The dynamic response of reef islands to sea-level rise: Evidence from multi-decadal analysis of island change in the Central Pacific. *Global and Planetary Change*, 72, 234–246.

Webster, P.J., Jian, J., Hopson, T.M., Hoyos, C.D., Agudelo, P.A., Chang, H.R., . . . Subbiah, A.R. (2010). Extended-range probabilistic forecasts of Ganges and Brahmaputra floods in Bangladesh. *Bulletin of the American Meteorological Society*, 91, 1493–1514.

WHO. (1946). Preamble to the *Constitution of the World Health Organization* as Adopted by the International Health Conference, New York.

WHO. (2014). *Quantitative Risk Assessment of the Effects of Climate Change on Selected Causes of Death, 2030s and 2050s*. World Health Organisation, Geneva.

WHO. (2015). *Environmental Health*. World Health Organisation, Geneva. http://www.who.int/topics/environmental_health/en (Accessed: 5 January 2015).

Wisner, B., Blaikie, P., Cannon, T. and Davis, I. (2004). *At Risk: Natural Hazards, People's Vulnerability and Disasters*. 2nd ed. Routledge, London.

Wisner, B., Gaillard, J.C. and Kelman, I., eds. (2012). *Handbook of Hazards and Disaster Risk Reduction*. Routledge, Abingdon.

Further readings

Akhtar, R. and Kovats, S. (2008). Climate, climate change and human health in Asian cities. *Environment and Urbanization*, 20(1), 165–175.

Campbell-Lendrum, D. and Corvalán, C. (2007). Climate change and developing-country cities: Implications for environmental health and equity. *Journal of Urban Health*, 84(1) Supplement, 109–117.

Forman, S., Hungerford, N., Yamakawa, M., Yanase, T., Tsai, H.J., Joo, Y.S., . . . Nha, J.J. (2008). Climate change impacts and risks for animal health in Asia. *Revue scientifique et technique*, 27(2), 581–597.

Haines, A. and Patz, J.A. (2004). Health effects of climate change. *JAMA: The Journal of the American Medical Association*, 291(1), 99–103.

Mol, L. and Sternberg, T., eds. (2012). *Changing Deserts: Integrating People and Their Environment*. White Horse Press, Cambridge.

Woodward, A., Hales, S. and Weinstein, P. (1998). Climate change and human health in the Asia Pacific region: Who will be most vulnerable? *Climate Research*, 11, 31–38.

5 Evolving a multi-hazard focused approach for arid Eurasia

Masato Shinoda

Introduction

Drylands occupy 41 per cent of the earth's landmass and are home to more than two billion people – a third of the world population in 2000. Most of these people are poor, located at the bottom of the economic pyramid. Climate projection studies have shown that in many regions, the frequency of extreme climate events is increasing, and that this trend will continue (IPCC 2013). This poses a particular threat to inhabitants of drylands whose livelihoods depend on agriculture and livestock rearing. This nature-society interactional system is among the most vulnerable to the projected increase in frequency of various kinds of extreme weather events.

Drylands at middle-high latitudes (such as those in Eurasia) have a harsh environment, coupled with a cold, arid climate (Figure 5.1). The livelihoods of people inhabiting these areas have long been jeopardised by repeated natural hazards associated with this climate. This area is regarded as a frontier (Anökumene) where the arid climate only sustains herding and/or hunting but not dry farming, and, furthermore, where the cold climate poses a considerable threat to the survival of livestock and wild animals during the freezing winter. In the natural setting of inland Eurasia with its cold, arid climate, meteorological (climatic) hazards are major threats, whereas geological hazards are secondary. For example, over the last century, destructive earthquakes occurred twice in Mongolia: in 1905 (Baljinnyam et al. 1993) and in 1957 (Kurushin et al. 1997). However, they were confined to the sparsely populated western and northern mountainous areas.

This chapter focuses on the landlocked Mongolian Plateau in the eastern part of the Eurasian continent (extending from Mongolia to China's Inner Mongolia). The Plateau lies at relatively high altitudes averaging 1,580m at middle latitudes. This geographical location is associated with a typical dry continental climate and cold winter. Hazards in Mongolia are characterised by the '4Ds': drought, *dzud* (harsh, cold-season conditions), dust storms and desertification that occur interactively. In this interaction, drought is a background factor that influences dust storms and dzuds, and the accumulated effect of multi-year droughts leads to desertification. Herein, drought memory is defined as a function in which below-normal anomalies of precipitation are carried over and maintained as anomalies of the earth surface system (Shinoda 2012), acting as a key mechanism of occurrences of the other hazards as explained later. However, these were not fully integrated into previous

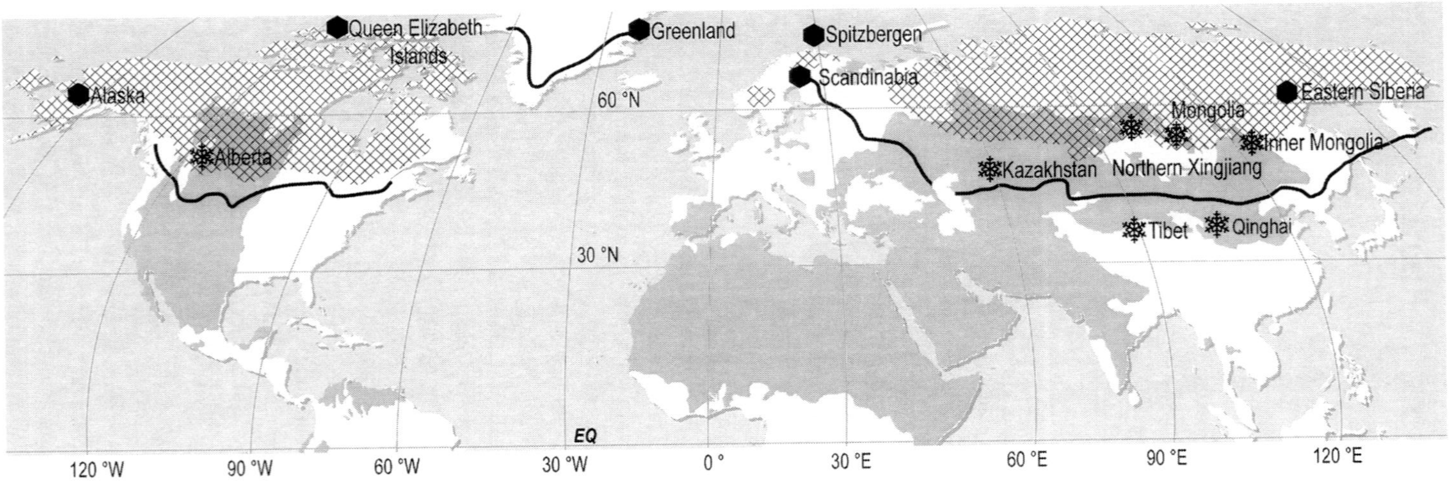

Figure 5.1 Northern Hemispheric distribution of cold, arid climates and documented *dzud*-like (snow symbols) and rain-on-snow (hexagonal symbols) phenomena. Mongolia is located within a unique climatic zone. The shaded area denotes drylands, while the hatched area and solid line indicate annual minimum temperatures below -40°C and -30°C, respectively

Source: Shinoda 2016

efforts to elucidate disaster mechanisms and implement appropriate land management techniques, as these have typically focused only on individual disasters. The livelihoods of herders who seasonally migrate across Eurasian pasturelands are affected not just by a single extreme weather event, but also by sequential events of the same or different extreme weather types. Although their accumulated effects harm herders' livelihoods and society, scientific research disciplines have paid little attention to these interactions.

Recently, a new approach, focusing on multiple hazards, has been evolving within disaster science (ARMONIA 2007; Kappes et al. 2012; Gill and Malamud 2014). The call for 'complete multi-hazard research' as a component of human settlement planning and management in disaster-prone areas originated within the UN Agenda 21 for sustainable development (UNEP 1992a). It was repeated in the Johannesburg Plan (UN 2002) and again in the Hyogo Framework of Action (UN-ISDR 2005). Given this context, disaster management authorities, responsible for making countermeasure decisions for specific geographic areas, must consider all spatially relevant hazards. They cannot reduce their focus to only one or two hazards like floods and landslides. This is because disaster management pertains to a particular spatial area (where the sum of hazards defines the overall spatial risk) and not to a particular object (as in the case of utility-providing sectoral planning institutions). Consequently, disaster management should ideally be based on a multi-hazard or multi-risk approach, referred to as 'all hazards at a place' (Hewitt and Burton 1971) or 'the totality of relevant hazards in a defined area' (Kappes 2011), to be able to appropriately handle risks and hazards in this spatial context. However, multi-hazard studies have been hindered by the strict separation maintained between scientific disciplines (WMO 1999).

Awareness of an increase in various types of extreme weathers that are likely to continue (IPCC 2013), and the crucial absence of a multi-hazard approach, motivated us to seek to understand the interactive nature of multi-hazards and to evolve an interdisciplinary approach, not only for establishing a disaster management framework, but also for its practical application in management. This chapter first identifies climate hazard dynamics in Mongolia such as dust storms, *dzud*s and droughts and their interactions. It then proposes a methodology for estimating the risk of hazards and developing an early warning system (EWS) of meteorological disasters for their proactive and integrated management.

Drought

Drought is the most hazardous natural disaster, worldwide, in terms of numbers of humans killed and/or adversely affected (Obasi 1994; Guha-Sapir et al. 2004). The geographic spread of its effects, which varies during its development and decline, is greater than that of other natural hazards (Wilhite 2000). A steep drying trend has been observed in soil moisture index values for land areas in the Northern Hemisphere since the mid-1950s, especially across northern Africa, Canada, Alaska and Eurasia, including Mongolia (Dai et al. 2004; Nandintsetseg and Shinoda 2014). In particular, below-normal precipitation in the Northern Hemisphere during 1999–2002 appears

to have extensively decreased vegetation activity over Eurasia and North America, as revealed by the satellite-estimated normalised difference vegetation index (NDVI) (Lotsch et al. 2005). Shinoda et al. (2014) used new indices of vegetation response (sensitivity and resilience) to drought to identify hotspots on the Asian Steppe that were vulnerable to the extensive four-year drought. In Mongolia, growing drought frequency has had increasingly important effects on animal husbandry and pasturing (Natsgadorj 2003; Nandintsetseg and Shinoda 2013).

Three types of drought have been defined in the Mongolian context: meteorological, agricultural and hydrological (Shinoda and Morinaga 2005). Meteorological drought is defined as a long-term anomalous deficit in precipitation; agricultural drought as an anomalous deficit in soil moisture in the root zone layer resulting from decreased precipitation and increased temperature (evapotranspiration); and hydrological drought as an anomalous deficit in surface water and groundwater. The timescales of these phenomena increase, chronologically, from the first to the last. These definitions are similar to those of Wilhite (2000).

Meteorological drought is usually the first stage of drought propagation throughout the entire hydrological cycle, and is, therefore, vitally important to monitor. The Standardized Precipitation Index (SPI) is widely used as a meteorological drought index (McKee et al. 1993), as this index is also used in Mongolia (Sternberg et al. 2011), while the Palmer Drought Severity Index (PDSI) is among the most widely used indices of agricultural drought (Palmer 1965). However, PDSI has several limitations, performing poorly as an indicator of short-term changes in soil moisture (e.g. Alley 1984; Guttman 1991), and discounting snowfall (Karl 1986). It may not, therefore, be suitable for monitoring agricultural drought in the cold climate that prevails at middle to high latitudes, including in Mongolia.

Agricultural (hereafter pasture) drought detection, and its monitoring index, have received less attention compared with meteorological drought indices, despite the negative impacts of this type of drought on the Mongolian Steppe. Soil moisture is appropriate for inclusion in a pasture drought index, because it reflects recent precipitation and antecedent stored water indicating agricultural potential (Keyantash and Dracup 2002). Moreover, soil moisture status in the root zone is a determining factor of vegetation growth (Miyazaki et al. 2004; Nakano et al. 2008; Nakano and Shinoda 2010; Shinoda et al. 2010b; Bat-Oyun et al. 2012) via water availability for transpiration. Consequently, Nandintsetseg and Shinoda (2011) modified a simple soil moisture model, intended for a wide range of practical applications, incorporating soil freezing and snowmelt to monitor pasture drought across Mongolia. Soil moisture values have been applied as a direct parameter of pasture drought in the following studies on climate change.

The first comprehensive assessment of meteorological and agricultural drought across Mongolia was conducted by Nandintsetseg and Shinoda (2013). They documented the frequency, duration and severity of growing-season (April–August) droughts, and their impacts on pasture productivity in Mongolia during 1965–2010. Meteorological and pasture drought characteristics were explored using SPI, the soil moisture anomalies percentile index and PDSI, applying a one-month timescale. All three indices showed a slight increase in the frequency and severity of droughts over this 46-year period, with significant dry conditions during the last

decade (2001–2010) in the four vegetation zones. Greater severity and frequency of growing-season droughts during the last decade have driven the reduction in pasture production in Mongolia. Furthermore, multi-decadal soil moisture trends in Mongolia were explored by Nandintsetseg and Shinoda (2014). They found a decreasing trend in soil moisture in all the vegetation zones from 1961–2006 resulting from decreased precipitation and increased potential evapotranspiration. However, the drying trend was only significant in the forest steppe. A process-based vegetation model, necessitated by adjusted climate projection data, showed a robust increase in the future August leaf area index, which is the indicator of summer drought and also the most important variable in estimating livestock loss caused by *dzud*s (Tachiiri and Shinoda 2012a). However, conducting a more spatially detailed examination of regional-scale drought, simulated by the global model, remains a challenge.

Dzud

In Mongolia, with its cold, arid climate, approximately 35 per cent of the workforce was engaged in stock farming in 2011. People living in rural areas are subjected not only to summer drought, but also to another climate hazard in the winter. *Dzud* is a Mongolian word indicating harsh winter conditions. Based on a detailed discussion of its definition (Natsagdorj and Dulamsuren 2001), *dzud* is defined, biogeophysically, as anomalous climatic and/or land surface (i.e. snow/ice cover and lack of pasture) conditions that lead to reduced accessibility and/or availability of pastures, and ultimately to significant livestock mortality during winter–spring. *Dzud* occurs throughout central Asia (Figure 5.1; Natsagdorj and Dulamsuren 2001; Hao et al. 2002; Robinson and Milner-Gulland 2003; Liang et al. 2007; Shang et al. 2012; Wang et al. 2013). In North America, a similar event termed *winter kill* occurred in the cold, arid plains of Canada's northern interior during the late 19th century (MacLachlan 2006).

While documentation of *dzud* events dates back 2000 years (Natsagdorj and Dulamsuren 2001), Tsedevsuren et al. (1997) have catalogued drought and *dzud* occurrences in central and eastern Mongolia from 1740 to 1921. Since the 1940s, the Statistical Office of Mongolia has recorded livestock mortality, providing a basis to identify a *dzud* event. Based on a synthesis of previous studies on *dzud*, Komiyama (2005) identified the following major *dzud* years: 1945, 1968, 1977, 2000, 2001 and 2002. Komiyama (2013) further quantitatively redefined and updated the post-1940 *dzud* inventory, based on the criterion of the mortality rate exceeding 10 per cent, and including the latest *dzud*, as follows: 1943, 1945, 1950, 1968, 2000, 2001, 2002 and 2010. The highest rate (33.2 per cent) was recorded for 1945, followed by 2010 (23.4 per cent).

In recent years, the underlying causes of *dzud* have been systematically analysed. Begzsuren et al. (2004) found that in southern Mongolia, livestock mortality resulting from a combination of growing-season drought and harsh winter weather was greater than that resulting from drought or *dzud* alone, or from neither of these conditions. Thus, in Mongolia, a disastrous year for livestock is likely to occur when a summer drought is followed by severe winter conditions. A similar drought–*dzud* linkage was observed in North America where periodic climatic

calamities caused by both drought and *winter kill* led to dramatic cattle losses from south of Alberta to western Texas during the frigid winter of 1886–1887 (MacLachlan 2006). Conversely, Sternberg et al. (2009) presented a view that on a multi-decadal scale, drought is not a determining factor for livestock loss in the Mongolian South Gobi desert.

Natsagdorj and Dulamsuren (2001) provided a conventional classification summary of *dzud* types in terms of direct factors contributing to conditions that prevent animals from grazing for consecutive days, finally resulting in their starvation. Referring to this pioneering work, a schematic classification of *dzud* types within a three-dimensional space and using three *dzud*-producing factors (axes) – snow/ice cover, lack of pasture and stormy weather – was presented in Shinoda and Morinaga (2005). White *dzud* is characterised by a combination of the first two factors and defined as conditions during which grasses that grow during the summer and decay during the subsequent cold season are covered by deep snow, preventing grazing. The snow depth during these conditions greatly exceeds plant height. This is the most common and disastrous *dzud* type. Similarly, iron (or glass) *dzud* is related to the first two factors, with grasses being covered with impenetrable ice that is produced through melt and refrozen snow (most likely occurring during spring and autumn). A similar phenomenon was observed in a rain-on-snow event in the north polar region that created ice layers at the surface of, within, or below the snow pack, killing the herd (Putkonen and Roe 2003). Hoof *dzud* is primarily associated with lack of pasture (the second factor), often caused by overgrazing. This may occur when a number of animals migrate to and are concentrated in relatively good but limited pasturelands during drought years. There are different definitions of black *dzud*, but their common feature is a lack of drinking water. This type of *dzud* is located at the low end (small amount of snow/ice) of the 'snow/ice cover' axis opposite the white *dzud*. This is because animals imbibe water through snow during very cold winters when water is frozen. Storms and cold *dzud*s both tend to result from stormy weather (the last factor), including strong winds, cold surges and/or dust outbreaks. These weather patterns reduce animals' intake of pasture that is determined by the availability of phytomass (i.e. not covered by snow and ice) and grazing time. Moreover, a series of the aforementioned *dzud* types during a single cold season is termed a multiple *dzud*. Multi-year *dzud*s like those that occurred during the three consecutive cold seasons from 1999/2000 to 2001/2002 may have an accumulated impact on pasture and livestock conditions.

While various *dzud* types have been documented within different bodies of literature, an attempt should be made to systematically and quantitatively measure *dzud* phenomena and their evolution. This can be done by assessing and predicting livestock weight, which most directly indicates an animal's health status and energy budget, as a combined result of *dzud*-causing factors. It is clearly difficult to accurately predict the effects of *dzud* on human lives and public health. Otani et al. (2014) have indicated that declining livestock numbers are associated with public health parameters such as infant mortality. Livestock losses may cause economic losses and a deterioration of herders' nutritional status. Thus, it is necessary

to conduct epidemiological surveys on disorders associated with these events, and to devise future countermeasures.

In Mongolia, pastoral livestock husbandry has suffered repeatedly from drought and *dzud*. *Dzud* occurrences between 1999–2002, during three cold seasons and preceding summer droughts, killed 11.2 million livestock or about 33 per cent of the total number of livestock in 1999. Between 2009 and 2010, 10.3 million heads perished because of *dzud*. These disasters impacted dramatically on national level economic conditions (Sternberg 2010), as clearly evidenced by the annual economic growth rate (Figure 5.2). During the most recent *dzud* event, colder-than-normal conditions prevailed in conjunction with easterlies over northern Mongolia (Koike et al. 2014). This is because the intensity of the *dzud*-producing cold surge was greater when the meandering of westerlies was enhanced and maintained north of Mongolia (with easterlies appearing over northern Mongolia). This synoptic pattern brought a northern cold air mass into this region. By contrast, the meandering was not strong at the end of November 2009, leading to near normal temperatures. In addition, earlier snow accumulation during November–December 2009 effectively strengthened radiative cooling on the ground surface, which frequently extends from Siberia to Mongolia just after the occurrence of cold air anomalies in an ambient atmosphere (Iijima et al. 2017). Another cold surge mechanism is the wave train pattern in the Northern Hemisphere that facilitates eastward migration of a cold air mass along with trough movement (Hori et al. 2011). The recent post-2000 winter climate over Eurasia has been characterised by frequent cold air anomalies in relation to the reduced extent of Arctic Sea ice and climate changes (Honda et al. 2009; Inoue et al. 2012). Historically, during

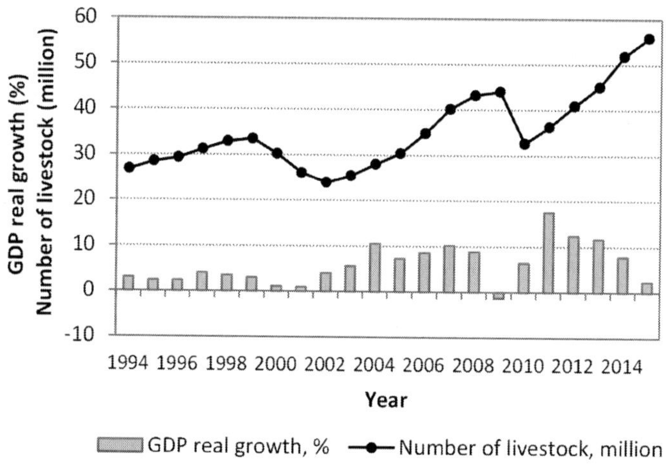

Figure 5.2 Real GDP growth rate (%) and number of livestock in Mongolia

Source: Shinoda (2015). doi:10.1016/j.piutam.2015.06.008. Reprinted with permission under the terms of the Creative Commons Attribution-NonCommercial-No Derivatives License (CC BY NC ND; https://creativecommons.org/licenses/by-nc-nd/4.0/)

the 1960s and 1970s, (the analysis period being 1960–2007), deep snow winters coincided with extreme cold, whereas beginning in 1992/1993, a new type of deep snow winter with warmer conditions occurred during some years (Koike et al. 2010). Koike and Shinoda (2011) carried out a daily-based analysis (1961–2007) of sharp, *dzud*-producing temperature drops after precipitation events, and their relationship to a cold front's passage.

Extended drought and *dzud* memory

Continental climates are established as a result of a complex interplay between the atmosphere and various land surface systems such as the biosphere, soil, hydrosphere and cryosphere. These systems function as climate memory, enabling the maintenance of seasonal or interannual (i.e. year-to-year) atmospheric anomalies. Climate memory is a function in which seasonal or interannual atmospheric anomalies (temperature, moisture and precipitation) are carried over and maintained as anomalies of the earth surface system. Seasonal anomalies indicate seasonal deviations from the average value over a year (or years), or the annual mean, while interannual anomalies denote seasonal deviations from the corresponding seasonal values averaged over multiple years. Climate memory is categorised according to component sub-systems of the entire climate system into atmospheric, vegetation and soil memories, based on the anomalies' characteristics: wet vs dry (drought) memories, hot vs cold memories, mild vs severe winter (*dzud*) memory etc.

Drought is a creeping phenomenon having a time-lagged carryover of anomalies in rainfall-soil moisture-pasture-livestock conditions and eventually culminating in a *dzud* (Shinoda and Morinaga 2005). That is, lower-than-normal precipitation ultimately leads to time-lagged, lower-than-normal livestock weight (health conditions) through the processes illustrated in Figure 5.3. The mechanism's chain starts from anomalies of atmospheric circulation (climate changes and variations) and resultant suppressed cyclone activities. These processes are referred to as drought memory. Here, drought memory is extended downstream to human livelihoods and public health, with the addition of the white and cold *dzud* memory flow and dust emission effects on human activity, as shown in Figure 5.3. Mortality is arguably the most reliable comparative indicator of human loss at the global scale. However, it is difficult to ascribe causes of mortality simply to drought as a natural event, because there are complex interactions between drought, political violence, chronic disease and economic poverty (Pelling 2014). Consequently, using the memory approach to examine drought's propagation effect would enable us to distinguish between drought- and non-drought-related mortality.

Shinoda and Nandintsetseg (2011) have demonstrated an interseasonal moisture memory mechanism mediated by land surface that is manifested in the cold, arid climate of Mongolia. Autocorrelation analysis of decay timescale (i.e. the lag at which autocorrelation function equals 1/e) revealed that in the Mongolian Steppe, soil moisture memory scales during the autumn and winter (8.2–5.5 months) are longer than those during the spring and summer (1.5–3.4 months). Low evapotranspiration and strong soil freezing, associated with the cold-season climate,

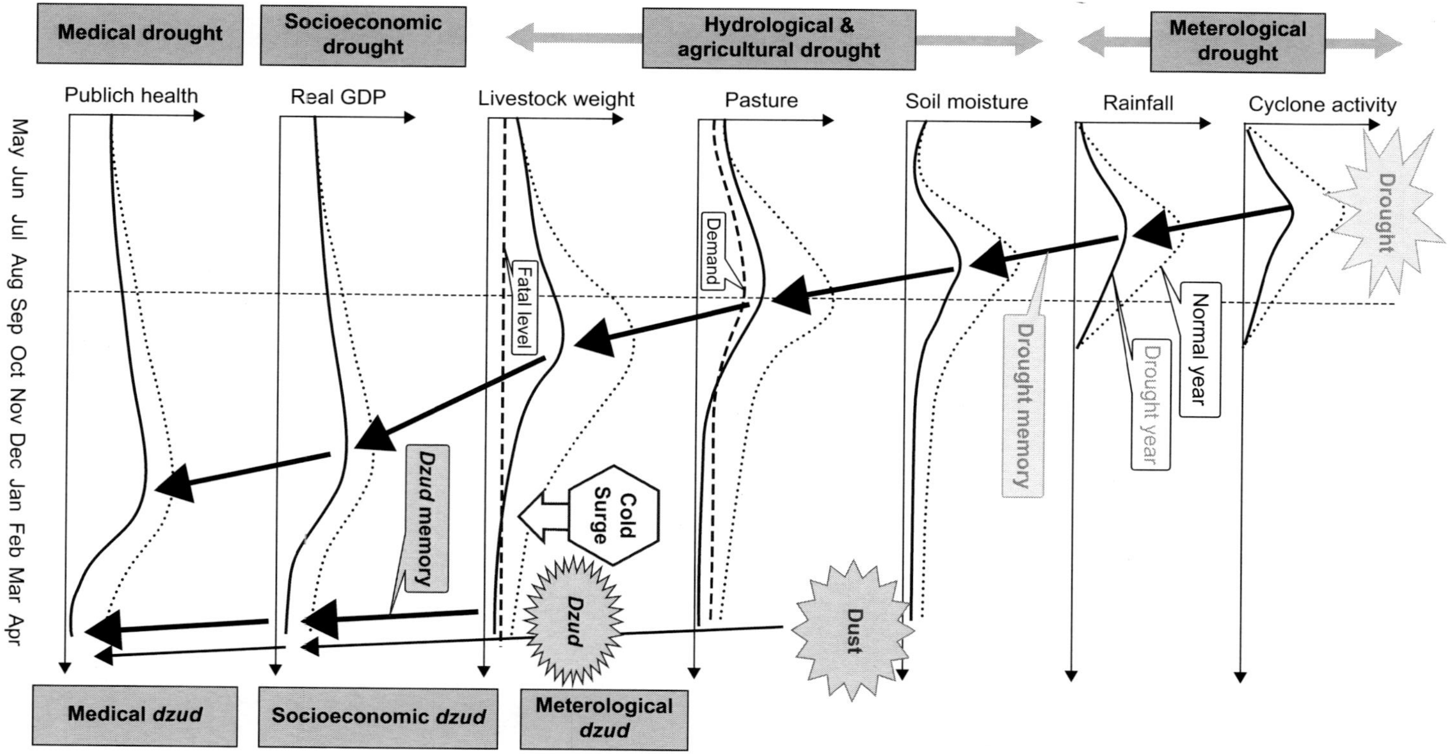

Figure 5.3 Drought and *dzud* memory processes through which hazards affect livestock and human health

Source: Shinoda (2015). doi:10.1016/j.piutam.2015.06.008. Reprinted with permission under the terms of the Creative Commons Attribution-NonCommercial-No Derivatives License (CC BY NC ND; https://creativecommons.org/licenses/by-nc-nd/4.0/)

prolong the decay timescale of autumn soil moisture anomalies to 8.2 months, which is among the longest timescales across the world. Vegetation memory exhibits a similar timescale (Shinoda and Nandintsetseg 2011). This is probably because the large rootstock of the dominant perennial plants in the Mongolian Steppe may survive, retaining belowground biomass anomalies during the winter (Shinoda et al. 2010b). This would then impact on initial vegetation growth during the spring.

Climate memory functions differently depending on the regional climate and land surface conditions. For example, the timescale of soil moisture memory differs across arid regions of the world: Mongolia, central Eurasia and the Sahel (south of the Sahara Desert). This memory during the spring (1.5 months) in Mongolia is almost the same as that observed in the cold arid region of central Eurasia, which is located at similar latitudes to those of Mongolia, but has deeper snow pack and shallower soil freezing (Shinoda 2001). The drying-up period after the rainy season is much longer in Mongolia (8.2 months) than that of soil moisture in the root zone observed in the tropical semi-arid Sahel (1.5 months) (Shinoda and Yamaguchi 2003). It is also considerably longer than the general decay timescale of two to three months (related to atmospheric forcing) for soil moisture in the top 1 m level, reported in the extratropics (Vinnikov et al. 1996; Entin et al. 2000).

To simulate vegetation responses to drought, and to better understand drought memory dynamics, a drought experiment (DREX) was conducted in the Mongolian Steppe during the rainy summer growing season of 2005 (Shinoda et al. 2010b). The manufactured drought drastically reduced aboveground phytomass (AGP) and soil water, but did not substantially affect belowground phytomass (BGP). AGP recovered quickly in the late summer of 2006, a likely reason being that BGP (which was several times greater than AGP) was not severely damaged by the drought. However, the poorly resilient species did not recover to pre-drought levels, suggesting that the response timescales differed among species. Despite intense drought, the large rootstock provided a basis for the quick recovery of AGP to pre-drought levels. In this regard, Nandintsetseg et al. (2010) have suggested that vegetation activity is primarily controlled by the current year's soil moisture and is slightly affected by underground structures stored in the root system. This research has been extended to the impact of drought on plant phenology (Shinoda et al. 2007) and diversity (Cheng et al. 2011) in Mongolian grasslands.

The Dust-Vegetation Interaction Experiment (DUVEX), a collaborative Japanese, Mongolian and American project, has played a significant role in revealing the relationship between drought memory and spring dust emission, and has contributed to the development of a temperate grassland wind erosion modelling system (Shinoda et al. 2011). Specifically, Kurosaki et al. (2011a, 2011b) have posited the dead-leaf hypothesis that the dead leaves of grasses in the spring, which are the residues of vegetation from the preceding summer, suppress dust outbreaks. In this context, a comprehensive understanding of drought memory dynamics has been achieved through a simulation of land surface conditions (Nandintsetseg and

Shinoda 2015). That is, soil moisture and vegetation memories, and the effects of grazing on grasslands, were realistically simulated using the DAYCENT ecosystem model (Kelly et al. 2000). This approach will be extended to integrate the DAYCENT vegetation growth and nutrient cycle models into a wind erosion model to investigate the sustainability of soil productivity under assumed conditions of future land use change and global warming (Nandintsetseg et al. 2014).

Drought memory research is now advanced regarding the downstream edge in relation to wild animals. Mongolia's Gobi-Steppe ecosystem, where substantial populations of migratory ungulates such as Mongolian gazelles (*Procapra gutturosa*) and Asiatic wild asses (*Equus hemionus*) are found, is the largest remaining area of intact steppe in the world, and, hence, of global importance (Batsaikhan et al. 2014). Kaczensky et al. (2010) have used satellite tracking to analyse the relationships between environmental conditions and ungulates' movements. This analysis suggests that they prefer higher NDVI areas in a seasonal (summer and winter) migratory range that varies on an interseasonal (Ito et al. 2006) and interannual (Ito et al. 2013) basis, implying that drought and *dzud* may influence habitat selection among wild ungulates.

In brief, the extended drought and *dzud* memory is a key concept for understanding how those hazards impact socio-economic activities and public health (Figure 5.3) and to warn of these events for strengthening herder preparedness. This has been attempted in Mongolia since the 2000–2002 *dzud* events, with the goal of creating a systematic EWS.

Dust storms

Severe dust events are frequent in arid regions of East Asia, particularly in the Taklimakan and Gobi deserts and on the Loess Plateau. Aeolian dust produced in these regions increasingly affects human health and local ecosystems, not only in the source regions but also downwind. Temperate grasslands (TGs) surrounding the desert source regions are sensitive to climate change, and are significant or potentially significant dust sources. TG aeolian processes are unique in that the vegetation growth-decay cycle and weathering process associated with extreme temperature changes profoundly affect the occurrence and intensity of wind erosion and dust emission (Shinoda et al. 2011). Human activities such as animal husbandry or cultivation may also result in land degradation and enhanced wind erosion.

Severe dust storms pose a serious threat to grassland livestock, and sometimes to human lives. For example, 120,000 and 110,000 animals were killed in China during the dust storms that occurred on May 5, 1993 and April 14–16, 1998, respectively (Shao and Dong 2006). A severe storm with dust and snow that occurred in eastern Mongolia on May 26–27, 2008 resulted in record level damages, killing 52 people and 280,000 animals (Chimgee et al. 2010). Thus, the need for an EWS to ameliorate damages is apparent. Moreover, a high-impact dust storm episode in 2007 was reported by Jugder and Shinoda (2011).

An overview report of the dust storm that occurred on May 26–27, 2008 was provided by the Government of Mongolia and the United Nations Development

Programme (UNDP) (2008) as follows. The severe dust storm, accompanied by snow, attacked five provinces in eastern Mongolia. During the storm, the air temperature dropped to nearly 0°C, and the wind speed appeared to reach over 40m/sec. This storm revealed a distinct structure with combined snowy and dusty weather. This included cloudy areas (with rain or snow) related to the low pressure system moving southeastwards, and dusty areas accompanying a cold air outbreak after the cold front's passage. The immediate report (Government of Mongolia and UNDP 2008) suggested that human factors exacerbated the damages. These included the following: inadequate transmission of weather information to local end-users (governors and herders), and their own misunderstanding of the information and consequent lack of preparedness for the storm. This event motivated us to study the relationship between weather conditions, socio-economic backgrounds and the damage caused by the dust storm, and to address the question of why the damage was localised (Chimgee et al. 2010). For our multiple regression analysis, we selected not only direct weather factors such as wind speed and precipitation, but also contextual human factors such as the previous year's livestock loss rate and the ratio of herders to the rural population as predictors of the rate of livestock loss. A previous interview-based survey (Ozaki 2010) suggested that in Berh village in Batnorov County of Khenti Province, dense human and livestock populations, in conjunction with the strengthened snowy storm, may have resulted in the highest mortality within this confined area.

The public health research group at Tottori University has revealed that the impacts of such a severe dust storm included not only direct impairment, but also long-term health damage that has decreased the quality of life of herders (Mu et al. 2010, 2011, 2013). Nevertheless, no information has been provided on its impacts on the health of domestic animals in Mongolia. However, Kobayashi et al. (2014) have revealed the effects of sand dust particles on the respiratory organs, including the lungs and tracheobronchial lymph nodes, of sheep and goats exposed to severe sandstorms.

There are two sets of erosion controls for any process: the forces that liberate particles from the main soil mass (erosivity), and the susceptibility of a particular soil to loss of material (erodibility) (UNEP 1992b). Erosivity is controlled primarily by wind strength, while erodibility is linked to land surface conditions. Regarding wind erosivity, dust emission is directly affected by strong winds associated with activities of extratropical cyclones, and on a larger scale, climate change, including global warming (IPCC 2013) and intensified continental-scale drought (Dai et al. 1998; Lotsch et al. 2005). Conversely, soil erodibility in grasslands is complicated by winter snow and the vegetation growth cycle, in which residual dead, brown leaves, remaining from the previous summer growing season, are covered by snow during the winter and provide the initial conditions for controlling aeolian processes during the spring (Shinoda et al. 2011). These processes are suppressed as plants grow in conjunction with the seasonal evolution of the summer rainy season. In Mongolia, the dust season is confined to the period when the ground surface is dry and free from snow and vegetation cover is minimal. Soil moisture is usually lowest during the spring (Nandintsetseg and Shinoda 2011) as a

high evaporation rate results from strong winds and snowmelt is reduced because of the limited snow pack (Morinaga et al. 2003). Seasonal freeze–thaw action is another factor that weakens aggregate stability and thus increases soil erodibility (Bullock et al. 2001). Grasslands are likely to be subjected to grazing pressure all year round, which affects erodibility.

The essential part of assessing the risk of dust disasters is deriving the relationship between threshold wind speed and land surface conditions. To achieve this goal, DUVEX was launched in the Mongolian Steppe (Shinoda et al. 2010a). This project demonstrated that although vegetation cover was very low (7.2 per cent) and the grazed plant height was not high, the vegetation substantially raised the threshold speed for saltation (Shinoda et al. 2010a). In particular, three years of continuous severe droughts preceded the 2008 spring at this site, most likely impacting the spring vegetation not only quantitatively (e.g. in terms of phytomass), but also qualitatively (e.g. through species composition) (Shinoda et al. 2010a). Furthermore, Abulaiti et al. (2013) described a substantial increase in the threshold wind speed during a year of high vegetation cover (2009), compared with a year of low vegetation cover (2008). Matsushima et al. (2012a) initiated a novel attempt to estimate the threshold wind speed over a non-vegetated surface using subsurface soil moisture estimated from thermal inertia (Matsushima et al. 2012b). DUVEX has also been extended to include a new site for dust observation located at Tsogt Ovoo in the Mongolian desert steppe (Ishizuka et al. 2012; Abulaiti et al. 2014). Recently, an advanced dust monitoring network, including PM2.5, PM10 and lidar observations, and covering the entire country, has been initiated in Mongolia (Jugder and Shinoda 2011; Jugder et al. 2011, 2012, 2014). This approach will enable us to extend the established relationship to a larger spatial scale using satellite data (e.g. Kimura and Shinoda 2010; Kimura 2017) as well as wide spanning ground observations, and to derive a wind erodibility map that indicates the potential of dust occurrence, namely, the threshold wind speed.

We have already developed an algorithm using a numerical simulation model and remote sensing technology to estimate surface wetness for each land surface from the macro-scale to the microscale (Kimura 2007; Kimura et al. 2007; Mohamed et al. 2011; Kimura and Moriyama 2014). Conditions that facilitate dust outbreaks in the source regions are: NDVI <0.2, normalised surface soil water content <0.2 (Kimura et al. 2009; Kimura and Shinoda 2010; Kimura 2012a, 2012b; Kimura and Moriyama 2014). Regarding short grasses that dominate in the temperate steppe, the threshold of vegetation cover above which wind erosion is efficiently eliminated is 15–20 per cent (Buckerly 1987; Lancaster and Baas 1998; Kimura et al. 2009). These criteria will be useful in practical land use management such as controlled grazing and reforestation.

On a decadal scale, the frequency of dust outbreaks has increased, while the frequency of strong winds has either decreased or shown little change from the 1990s to the 2000s in Mongolia, eastern Inner Mongolia and Northeast China (Kurosaki et al. 2011b). In conjunction with this trend, threshold wind speed has decreased in these subregions. This suggests that changes in land surface conditions have increased vulnerability to dust outbreaks, resulting in their increased

frequency. In some areas of the Mongolian grasslands, decreased precipitation has caused a reduction in the amount of vegetation in summer, leading, ultimately, to a reduction in the amount of brown, dead leaves during the following spring. The reduction in dead leaves has increased dust emission through reductions in shear stress and protective cover. Recently, more attention has been paid to temperate grasslands (widespread over arid Eurasia) that are sensitive to climate change and are significant, or potentially significant, dust sources (Shinoda et al. 2011). In the cold, arid climate of Eurasia, interannual anomalies of soil moisture and vegetation resulting from the quantity of rainfall during a given summer are maintained throughout the freezing winter months until the spring (Shinoda and Nandintsetseg 2011). The concept of memories will enable us to predict dust emission conditions, a half year in advance, through detailed monitoring and assessment of the time span of the soil–vegetation system on the ground and by satellite.

Desertification

The Eurasian grasslands (or steppes), where nomadic pastoralism has been a major livelihood source for thousands of years, are facing desertification. Temperate grasslands (TGs) are situated in mid-latitude regions where climatic conditions favour the dominance of perennial grasses (Archibold 1995). The most extensive TGs are the prairies of North America, previously extending over 350 million ha, but now reduced to about 60 million ha. In Eurasia, the steppes encompass 250 million ha of rolling plains that extend from Hungary to Northeast China. TGs form a buffer zone between deserts and forests and can act as a frontier for the expansion of deserts or forests, depending on prevailing climatic conditions.

The most widely accepted definition of desertification, globally, is provided in the United Nations Convention to Combat Desertification (UNCCD). Here, it is defined as land degradation in (climatologically defined) arid, semi-arid and dry subhumid areas resulting from various factors, including climatic variations and human activities (UNEP 1992b). To provide more detail, this has two major aspects: soil and vegetation degradation. Causes of soil degradation include wind and water erosion and chemical, physical or biological degradation. Vegetation degradation may be caused by over cultivation, overgrazing and overharvesting. While this definition does not encompass the concept of timescale, based on which land degradation evolves and disappears, desertification is also defined as a temporary or permanent lowering of the productive capacity of land (UNEP 1992c), thus underscoring its reversible or irreversible nature. For example, if the impacts of water or wind erosion are sufficiently large, the degraded land would take thousands of years of non-use to recover its plant nutrients and organic matter. Irreversibility is a consideration in identifying hotspots on the Asian grasslands (Shinoda et al. 2014).

A global assessment of land degradation indicated that Northeast China's grassland area is a hotspot of wind and water erosion resulting from decreased vegetation cover (Lepers et al. 2005). A previous assessment for Mongolia suggested that 78.2 per cent of its territory has been affected by moderate to very severe desertification (mainly in terms of soil erosion) attributed to overgrazing (Mandakh et al.

2007). The most recent estimate of the proportion of the country's area affected by desertification (including soil and vegetation degradation) is 77.8 per cent, with the four defined degrees ranging from slight to very severe (Institute of Geo-ecology 2013). Regionally, on the Mongolian Plateau, water erosion is predominant in the forest–steppe area, wind erosion in the steppe–desert, while vegetation degradation affects the entire region.

TGs deserve particular attention from aeolian researchers, because they are subject to severe seasonal wind erosion activities that threaten the sustainability of their ecosystems (Shinoda et al. 2011). The sustainability of this grassland ecosystem critically depends on interactions between climate change, grassland vegetation and human activities. Aeolian processes play a major role in these interactions. For example, overgrazing may result in decreased vegetation and increased wind erosion that constrains the recovery of grasslands during the subsequent growing season. Moreover, a marked drying trend in eastern Eurasia's arid region during the past century (IPCC 2007) may have had an adverse impact roughly equivalent to that of overgrazing.

The dynamics of wind erosion in TGs evidence unique features. The most apparent among these is the major role that the vegetation growth-decay cycle plays in seasonal variations of wind erosion activities. For example, Mongolian grasslands generally experience severe wind erosion in the spring. This is probably because of the lack of aboveground phytomass (Shinoda et al. 2010a) and the occurrence of strong winds (Kurosaki and Mikami 2005). During the growing season, however, wind erosion is almost negligible. Aerodynamic characteristics and the protective effects of vegetation substantially differ during the growth and decay seasons. A vegetation cover of 15–20 per cent may be the approximate threshold that demarcates seasons as well as areas affected or not affected by wind erosion. TGs experience extensive seasonal variations in temperature across the freezing point. These changes promote soil weathering processes, particularly through freeze–thaw mechanisms, unlike dust source areas at lower latitudes.

Vegetation degradation drives declining pasture yields, which are likely, ultimately, to result in limited forage choices for animals and increasing incidents of plant poisoning and resultant animal losses (Damiran and Darambazar 2003). For example, the spread of poisonous plants including *Oxytropis glabra*, is a serious problem associated with such desertification in China's northwestern grasslands, which are adjacent to western Mongolia (Zhao et al. 2011). In desert areas of western Mongolia, a neurological disorder has also developed in goats, sheep, cattle and horses during the last five years, with resultant economic losses. Veterinarians have suggested that there is an association between this disorder and ingestion of *O. glabra* (Takeda et al. 2014).

Integrating dryland disaster science: A new approach

As discussed in the introduction, disasters in dry Eurasia can be characterised as the '4Ds': drought, *dzud*, dust storms and desertification that occur interactively.

However, previous attempts to elucidate disaster mechanisms and implement appropriate land management techniques have not been fully integrated, because these efforts have typically focused only on individual disasters. Against this background, a new project, 'Integrating Dryland Disaster Science' has been implemented for FY 2013–2017 under the Grants-in-Aid for Scientific Research Program supported by the Japan Society for the Promotion of Science. This project adopts a novel approach by exploring interactions between hazards in the context of drought memory.

The project's dual aims are interrelating 4D disasters in dry inland areas of Eurasia in terms of their causal mechanisms (especially drought memory) and the timescales of their occurrence, and developing comprehensive proactive countermeasures and making policy recommendations designed to mitigate multi-disaster impacts. The first objective can be achieved by correlating the impacts of extreme weather events producing the 4Ds in terms of the drought memory concept; evaluating risks for each disaster as the product of impact (hazard or forcing); and evaluating vulnerability (expressed through the three elements of exposure, sensitivity and resilience) of the nature-society system (Figure 5.4). Assessments of vulnerability should be made using socio-economic, public health and environmental data. The second objective can be achieved by implementing approaches that integrate the opinions of various stakeholders on countermeasures needed to reduce the vulnerability of both people and the environment to 4D hazards.

Vulnerability is a concept that evolved out of the social sciences and was introduced as a response to purely hazard-oriented predictions of disaster risk that prevailed in the 1970s (Schneiderbauer and Enrlich 2004). It is defined as conditions determined by physical, social, economic and environmental factors or processes that increase a community's susceptibility to the impact of hazards (UN-ISDR 2005). A similar definition was adopted by UNEP (2004). Thus, hazards and vulnerability are both regarded as crucial factors that determine risks. Although hazard-oriented research provided the initial motivation for developing the 4D project, evolving a vulnerability-oriented approach for understanding the overall structure and function of disaster dynamics (illustrated in Figure 5.4), and measuring societal and environmental vulnerability, remains a considerable challenge. This topic has provoked a wide range of discussions and debates from different perspectives that vary with scientific fields and approaches (see Birkman 2014 for a comprehensive review). According to the review by Birkmann (2006), our approach correlates with the global environmental change conceptual model (Turner et al. 2003), or the pressure and release model (Wisner 2004). However, it is not viewed as a holistic approach that encompasses disaster risk management as found in the Bogardi/Birkmann/Cardona conceptual framework. Nevertheless, it is currently the only available ecology-based analytic, as discussed below.

In Figure 5.4, drought as a hazard (forcing) is linked to other hazards with different timescales by the drought memory mechanism (shown in Figure 5.3). Drought is a background factor that influences dust storms and *dzud*s, and the accumulated effect of multi-year droughts leads to desertification. Its risk (influence) is calculated as the product of hazard and vulnerability of the natural environment-society

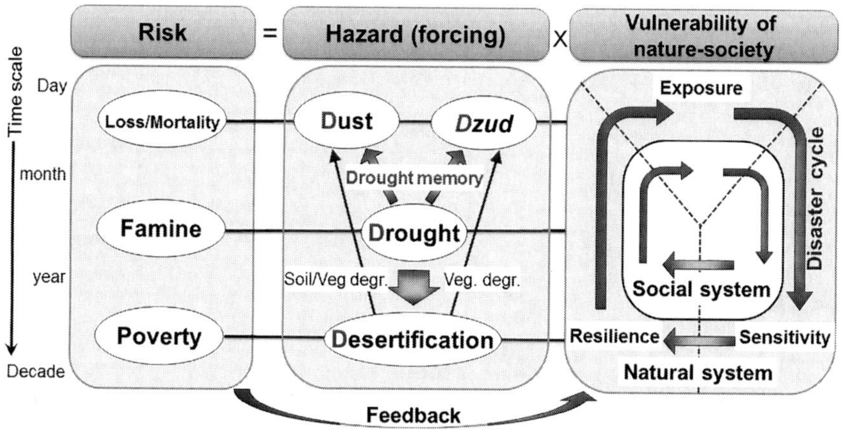

Figure 5.4 Interrelationships between 4D disasters (dust, *dzud*, drought and desertification) and the drought memory mechanism

Source: Shinoda (2015). doi:10.1016/j.piutam.2015.06.008. Reprinted with permission under the terms of the Creative Commons Attribution-NonCommercial-No Derivatives License (CC BY NC ND; https://creativecommons.org/licenses/by-nc-nd/4.0/)

system. In this conceptual model, each type of hazard is related to specific types of risk. Sporadic and intensive hazards such as dust storms and *dzud*s are related to economic and/or infrastructural losses and mortality, while drought has a longer timescale and desertification an even longer one, leading to famine and poverty, respectively. The vulnerability of both the natural world and of society is expressed by the three elements: exposure, sensitivity and resilience in a clockwise sequence of a disaster cycle involving the social and natural domains. Damages caused by disasters provide feedback relating to vulnerability.

Measuring vulnerability to multi-hazards

Adopting an ecology-based approach, vulnerability can be expressed by the three elements: exposure, sensitivity and resilience and extended to the herders' society. Previous ecological studies have viewed sensitivity as resistance (Tilman and Downing 1994; Pfisterer and Schmid 2002; Bai et al. 2004), while resilience has been viewed as recovery (Pfisterer and Schmid 2002). For example, in ecological studies, drought resistance was measured using the relative rate of biomass change between pre-drought and drought years (dB/B dt, yr^{-1}; Tilman and Downing 1994). In climate change studies, climate sensitivity has been similarly formulated as the relative rate of climate change to an external forcing's change (e.g. Schneider and Dickinson 1974; Charney et al. 1979).

Vulnerability is defined simply as a system that produces disaster (or risk) as an output, with hazard as an input. This definition is analogous to that used in

sensitivity analysis, which has been applied within various natural and social sciences (e.g. Saltelli et al. 2000). For example, a decrease in precipitation (drought event) as an input (ΔI_i) into event i is converted through the system with a property of sensitivity (S_i) to a change (likely decrease) in biomass as an output (ΔO_i). Similarly, an increase in precipitation (recovery from a drought event) as an input (ΔI_{i+1}) at event i+1 is converted through the system with a property of resilience (R_i) to a change (i.e. increase) in biomass as an output (ΔO_{i+1}). Hence, the characteristics of the conversion by the system can be used to characterise the system's response. A practical analysis deals with a dominant drought event with an amplitude of ΔI that is large enough (e.g. exceeding a threshold) to result in a detectable signal of ΔO, because the system has its own inertia to absorb the shock of the event (Shinoda et al. 2014).

The advantage of this method is that the simple ratio of ΔO/ΔI can be used to characterise a system as a black box. However, it is vital to carefully crosscheck between the index and an analysis of a system's vulnerability that encompasses physical, social, economic and environmental elements. Vulnerability analysis for Mongolia was conducted under the 4D project (Du et al. 2017a, 2017b). Middleton et al. (2015) have also implemented a similar approach for a western Mongolian province. Future studies should extend this single-hazard approach to include multiple simultaneous hazards and develop an analytic method for quantitative apportionment of the combined multi-hazard impacts with respect to their various forcings. This will enable the identification of high-impact hazards, which should be given high priority within disaster risk management practice. Moreover, a process-based model that simulates fattening or starving livestock is being developed to analyse the combined effect of multi-hazards and vulnerability of the social-natural system on them (Tachiiri and Shinoda 2014).

Proactive management

The Information and Research Institute of Meteorology, Hydrology and Environment (IRIMHE) in Mongolia (formerly the Institute of Meteorology and Hydrology) has been in charge of meteorological, agrometeorological and hydrological operations of observation and research, providing a firm basis for the monitoring and forecasting of drought and *dzud* events (see Shinoda and Morinaga 2005). In particular, drought and *dzud* indices, calculated using monthly precipitation and temperature data, have been developed and used mainly for research.

In general, a disaster management cycle encompasses the pre-disaster phases of mitigation/prevention and preparedness and the post-disaster phases of response and recovery. Preparedness for a *dzud* is ensured on account of the time lag between a drought in summer and a *dzud* in winter–spring. Monitoring and forecasts related to soil moisture and pasture are essential needs for a drought–*dzud* EWS. Thus, a regression tree model was developed to provide a forewarning of *dzud*. The predictor variables included remotely sensed vegetation and snow quantity data, as well as the previous year's livestock numbers and mortality (Tachiiri et al. 2008). Shinoda (2012) outlines four major *dzud*-producing factors that were

selected in the model. Background factors include livestock conditions during the previous year, that is, the condition of possibly weakened livestock that survived events leading to large (quantified) livestock losses or small carrying capacity based on measurements for many livestock. Environmental factors are related to poor pasture conditions in summer and deep snow in winter, both of which weaken livestock. If some of these factors occur in combination, *dzud* is more likely to occur. This model confirmed that the most serious livestock mortality in winter–spring was associated with poor vegetation during the previous year's summer, substantial snow in winter and a high mortality rate during the previous year. This information, which includes the disaster-inducing thresholds of each parameter for each time segment in a year, will be used to create guidelines for an EWS and a timeline for achieving preparedness. A similar tree regression model has been applied to future *dzud* risk assessment using projections of future climate changes (Tachiiri and Shinoda 2012b).

Given that vegetation conditions during the preceding summer are among the most crucial drivers of *dzud* (as indicated in the earlier analysis), pasture conditions in Mongolia have been mapped since 2001. This initiative was triggered launched during the three consecutive winter *dzud*s from 1999/2000 to 2001/2002. This effort was followed by the launching of the Project for Development of Human Capacity for Weather Forecasting and Data Analysis in Mongolia by the Japan International Cooperation Agency (JICA). The project has produced a map on a finer (village) spatial scale by collecting positional information for each GPS pasture measurement (see Shinoda 2012). Previously, locational information for the pasture data was not available, and the data in a county (an administrative unit larger than a village) was just averaged. The project has provided local observers with GPS receivers for collecting locational information. This map is advantageous in that more locations, albeit small, with good pasture conditions, can be identified on a village scale, compared with a county scale, even using the same dataset. Based on this map, herders can be guided to a good pastureland that is closer than the areas they previously used. Their movements are consequently dispersed to more areas, resulting in the mitigation of land use pressure on good pastures. This mapped information will be crucial for achieving optimised use of pasturelands to prevent localised land degradation resulting from high grazing pressure.

Traditional herding knowledge

As discussed in this chapter, Mongolia's climatological context poses challenges for the survival of herders. However, the occupation of herding, related to domesticated horses, in the Eurasian Steppe is thought to have survived for several thousands of years in the heart of the continent (Levine 1999; Hongo 2006). This history implies that herders have long retained indigenous knowledge and wisdom that has ensured the sustainability of their livelihoods. This relates to herding practices, environmental perceptions, pasture management and, most crucially, disaster management. Hereafter, this knowledge is referred to as traditional herding

knowledge (THK), covering the totality of the herders' knowledge produced from the inception of herding to the present, and encompassing traditional ecological knowledge (TEK) (Fernandez-Gimenez 2000). In relation to herding practices, region-specific seasonal migration patterns (Bazargur 2005) are at the core of THK. Concerning pasture management, Genghis Khan's instruction 'Do not spade the soil' can be scientifically interpreted. Spading exposes shallow soil on grasslands to strong winds, resulting in drying and erosion of the soil and ultimately creating sterile land. This is a typical case of irreversible land degradation, as described in the desertification section, which most crucially endangers herding sustainability.

Moreover, wise, old Mongolian adages, notably those preserved in the lessons of *Togtohör* or *To* (an abbreviation) *Wang* (meaning King i.e. King *Togtohör/To*), are thought to have documented portions of knowledge that relate to pre-disaster management in response to dramatic livestock losses from a combination of drought and *dzud* events that occurred around the middle of the 19th century (Hagihara 1999). Examples of these lessons include: 'a weather phenomenon with persistent easterlies during the winter is a precursor of a *dzud*,' and 'livestock should be fattened sufficiently through the early summer to survive the following severe winter season.' The first lesson suggests a synoptic weather pattern like a blocking high that persistently brings cold air from the north into this area, as indicated for contemporary *dzud* events by Koike et al. (2013). The second didactic saying appears to be scientifically sound based on a statistical analysis of livestock weight (Morinaga et al. 2004). In general, the weight of cows substantially increases owing to their grazing on summer grasses from May to September, whereas it decreases from October to April as a result of the cold weather and nutritionally poor grasses available for grazing. This analysis indicates a significant negative correlation between the rate of the increase and that of the decrease in weight, namely, enhanced fattening of animals during the early summer is doubly efficient to withstand the harsh winter. It implies evidence in favour of the second didactic saying cited earlier. Without such knowledge, the herders' society may not have survived for thousands of years as a result of extreme climate events that have repeatedly occurred throughout their history (Natsagdorj and Dulamsuren 2001). Scientific verification of such knowledge has just been initiated within our projects, focusing on *airag* (fermented mare's milk) production (Bat-Oyun et al. 2015), seasonal camp site selection and other relevant topics.

Applying a vulnerability approach to disaster preparedness

This chapter emphasises the dry stabilised (i.e. weak earthquake and volcanic activities) inland of Eurasia, in contrast with the wet mobile coastal belt as treated in the Natural Hazards Assessment Network world map of natural hazards (Munich Re 2011). As mentioned in the THK section, the herding occupation on the inland Eurasian Steppe is thought to have survived the aforementioned 4D hazards for several thousands of years. This history implies that in a changing climate (including hazards), herders have long maintained a sound herder-pastureland system, relying

on indigenous herding and ecological knowledge. A major body of this knowledge includes mobile pastoralism that avoids a concentration of grazing pressure in limited areas and thereby vegetation degradation. Moving herds in search of fresh pasture and water use pasturelands extensively but lightly. In this regard, vulnerability may be analysed by examining the stability of system response to repeated forcings (as mentioned in the section on measuring vulnerability to multi-hazards), which avoided historical collapse. These forcings include negative ones that lead to collapse – i.e. extreme weather, disease, interrupted commerce, immigration and war – and positive ones leading to stability – i.e. introduction of modern knowledge (science and technology), international aid and peace.

In the sensitivity analysis, the system vulnerability issue can be regarded simply as the system's response that produces disaster (or risk) as an output, with hazard as an input (forcings or perturbations), which is applicable to multi-hazards. The vulnerability can be expressed by the elements of exposure, sensitivity and resilience (Figure 5.4). For example, for the disastrous 2008 dust storm (see the section on dust storms), three stages related to each element were identified with regard to how the society responded to the storm. The first stage was associated with their exposure to the storm (e.g. strong winds and dust). If the people detected a storm prelude and found shelter or evacuated the affected area, they could avoid the physical impacts. If they instead got caught in the storm, they experienced the second stage, in which their sensitivity (conversely resistance) to the storm was tested to the extent that the storm impacted their physical conditions. Finally, when they experienced adverse storm impacts, the third stage occurred, with resilience (or recovery) toward normal conditions with self-help efforts and external aids. This analytic framework may be effective in examining human response to a hazard in the disaster cycle and relating it to human vulnerability.

Once we understand the nature of the system response to a single hazard, the three vulnerability elements are expressed as a function of environmental, health and socio-economic parameters, which are then integrated on a vulnerability map. As illustrated in the framework of risk assessment (Figure 5.4), this map is combined with each hazard map to produce risk maps for each hazard and, ultimately, to produce an integrated risk map. Countermeasure options for disaster preparedness may influence corresponding elements of vulnerability, and thus integrated risk. This approach provides a pre-disaster assessment of countermeasure effectiveness as a basis for policy recommendation.

Summary and challenges

With a focus on arid Eurasia, especially Mongolia, this chapter has outlined how a multi-hazard approach can be applied to natural disaster science and management. In brief, based on drought memory, 4D-related hazards are seen to be dynamically inter-related, and a deeper understanding of this linkage as well as the development of extreme weather forecasting will enable the creation of a holistic EWS encompassing all of the 4Ds. Bridging disaster and area studies, however, is a key challenge for area-based management.

Consequently, future challenges to be prioritised in the context of disaster management are as follows. First, a comprehensive 4D EWS (4DEWS) is crucial for developing proactive management. The already established pasture capacity map is key to this effort. However, near real time information regarding snow, livestock health and the environment should be incorporated to expand the system and create guidelines and a timeline for proactive forewarning of disasters. Process-based animal weight modelling, combined with a weather forecast, is the most efficient approach for including various environmental and socio-economic effects within a single parameter (i.e. weight), and to identify the fatal threshold of *dzud*-afflicted animals. Second, in the context of 4DEWS, a multi-hazard approach must integrate disaster studies within area studies. This is because local areas suffer the combined impacts of simultaneously occurring hazards, regardless of their origins and differing mechanisms. Timely and proper dissemination of readable 4DEWS information to local governors and herders is also vital. Third, the methodology of vulnerability analysis should be established on a detailed spatial scale (e.g. the county or village scale), using available socio-economic and environmental data (such as those adopted by Du et al. 2017b), to identify and prioritise hotspots for proactive action prior to a forewarning of hazard. This analysis will also help answer an important question: why were the 2009/2010 *dzud* impacts so serious within a specific region? It also provides lessons regarding local preparedness for future disasters. Finally, scientific verification of THK is key to practically applying this knowledge to pastures, herding and disaster management.

References

Abulaiti, A., Kimura, R. and Shinoda, M. (2013). Vegetation effects on saltation flux in a grassland of Mongolia. *Sand Dune Research*, 59(3), 117–128.

Abulaiti, A., Kimura, R., Shinoda, M., Mikami, M., Ishizuka, M., Yamada, Y., . . . Gantsetseg, B. (2014). An observational study of saltation and dust emission in a hotspot of Mongolia. *Aeolian Research*, 15, 169–176.

Alley, W. M. (1984). The palmer drought severity index: Limitations and assumptions. *Journal of Applied Meteorology and Climatology*, 23, 1100–1109.

Archibold, O. W. (1995). *Ecology of World Vegetation*. Chapman and Hall, London.

ARMONIA. (2007). Assessing and Mapping Multiple Risks for Spatial Planning. European Union 6th Framework Programme Reports. European Union.

Bai, Y., Han, X., Wu, J., Chen, Z. and Li, L. (2004). Ecosystem stability and compensatory effects in the Inner Mongolia grassland. *Nature*, 431, 181–183.

Baljinnyam, I., Bayasgalan, A., Borisov, B. A., Cisternas, A., Dem'yanovich, M. G., Ganbaatar, L., . . . Vashchilov, Y. Y. (1993). Ruptures of major earthquakes and active deformation in Mongolia and its surroundings. *Geological Society of America Memoir*, 181, 62.

Bat-Oyun, T., Erdenetsetseg, B., Shinoda, M., Ozaki, T. and Morinaga, Y. (2015). Who is making *airag* (Fermented Mare's Milk)? A nationwide survey on traditional food in Mongolia. *Nomadic Peoples*, 19(1), 7–29.

Bat-Oyun, T., Shinoda, M. and Tsubo, M. (2012). Effects of water and temperature stresses on radiation use efficiency in a semi-arid grassland. *Journal of Plant Interactions*, 7(3), 214–224.

Batsaikhan, N., Buuveibaatar, B., Chimed, B., Enkhtuya, O., Galbrakh, D., Ganbaatar, O., . . . Whitten, T. (2014). Conserving the world's finest grassland amidst ambitious national development. *Conservation Biology*, 28(6), 1736–1739.

Bazargur, D. (2005). *Geography of Pasture Animal Husbandry*. Institute of Geography, Mongolian Academy of Sciences, Ulaanbaatar.

Begzsuren, S., Ellis, J.E., Ojima, D.S., Coughenou, M.B. and Chuluun, T. (2004). Livestock responses to drought and severe winter weather in the Gobi Three Beauty National Park, Mongolia. *Journal of Arid Environments*, 59, 786–796.

Birkmann, J. (2006). *Measuring Vulnerability to Natural Hazards: Towards Disaster Resilient Societies*. United Nations University Press, Tokyo.

Buckerly, R. (1987). The effect of sparse vegetation in the transport of sand dune by wind. *Nature*, 325, 426–428.

Bullock, M.S., Larney, F.J., Izaurralde, R.C. and Fenp, Y. (2001). Overwinter changes in-wind erodibility of clay loam soils in Southern Alberta. *Soil Science Society America Journal*, 65, 423–430.

Charney, J.G., Arakawa, A., Baker, D.J., Bolin, B., Dickinson, R.E., Goody, R.M., . . . Wunsch, C.I. (1979). *Carbon Dioxide and Climate: A Scientific Assessment*. National Academy of Sciences, Washington, DC. 22.

Cheng, Y., Tsubo, M., Ito, T., Nishihara, E. and Shinoda, M. (2011). Impact of rainfall variability on plant diversity in grazing land: A case study from Mongolia. *Journal of Arid Environments*, 75(5), 471–476.

Chimgee, D., Shinoda, M., Tachiiri, K. and Kurosaki, Y. (2010). Why did a synoptic storm cause a dramatic damage in a limited area of Mongolia? *Mongolian Population Journal*, 19, 63–68.

Dai, A., Trenberth, K.E. and Karl, T.R. (1998). Global variations in droughts and wet spells: 1900–1995. *Geophysical Research Letters*, 25, 3367–3370.

Dai, A., Trenberth, K.E. and Qian, T. (2004). A global data set of Palmer drought severity index for 1870–2002: Relationship with soil moisture and effects of surface warming. *Journal of Hydrometeorology*, 5, 1117–1130.

Damiran, D. and Darambazar, E. (2003). *Toxic Plants of Mongolian Rangelands*. Oregon State University, Eastern Oregon Agricultural Research Station, Union, OR, U.S.A.

Du, C., Shinoda, M., Tachiiri, K., Nandintsetseg, B., Komiyama, H. and Matsushita, S. (2017a). Mongolian herders' vulnerability to *dzud*: A study of record livestock mortality levels during the severe 2009/2010 winter. Submitted to *Natural Hazards*.

Du, C., Shinoda, M., Komiyama, H., Ozaki, T. and Suzuki, K. (2017b). Social factors accounting for local differences in the damage caused by a meteorological disaster: A case study of the 2009–2010 *dzud* in Mongolia. *Journal of Arid Land Studies*, in press (In Japanese with English abstract).

Du, C., Komiyama, H., Shinoda, M., Matsushita, S. and Tahiri, K. (2014). Social Vulnerability Analysis of Dzud Disaster in Mongolia. *International symposium on multi-hazard approach in Mongolia*. Ulaan Baatar.

Entin, J.K., Robock, A., Vinnikov, K.Y., Hollinger, S.E., Liu, S. and Namkai, A. (2000). Temporal and spatial scales of observed soil moisture variations in the extratropics. *Journal of Geophysical Research*, 105, 11865–11877.

Fernandez-Gimenez, M.E. (2000). The role of Mongolian nomadic pastoralists' ecological knowledge in rangeland management. *Ecological Applications*, 10, 1318–1326.

Gill, J.C. and Malamud, B.D. (2014). Reviewing and visualizing the interactions of natural hazards. *Reviews of Geophysics*, 52, 680–722.

Guha-Sapir, D., Hargitt, D. and Hoyois, P. (2004). *Thirty Years of Natural Disasters, 1974–2003: The Numbers*. Presses Universitaires de Louvain, Louvain-la-Neuve, Belgium.

Guttman, N. B. (1991). A sensitivity analysis of the Palmer hydrologic drought index. *Water Resources Bulletin*, 27, 797–807.

Hagihara, M. (1999). Paradigms of the Mongol study: About the teachings of TO-WANG: Teaching about the nomadic life in Khalkha-Mongolia during the 19th century. *Bulletin of the National Museum of Ethnology*, 20, 213–285.

Hao, L., Wang, J., Man, S. and Yang, C. (2002). Spatio-temporal change of snow disaster and analysis of vulnerability of animal husbandry in China. *Journal of Natural Disasters*, 11, 43–49.

Hewitt, K. and Burton, I. (1971). *Hazardousness of a Place: A Regional Ecology of Damaging Events*. Toronto Press, Toronto.

Honda, M., Inoue, J. and Yamane, S. (2009). Influence of low Arctic sea-ice minima on anomalously cold Eurasian winters. *Geophysical Research Letters*, 36, L08707.

Hongo, K. (2006). Domestication of dromedary and its propagation. *Bulletin of the Society for Western and Southern Asiatic Studies*, 65, 56–72 (In Japanese).

Hori, M. E., Inoue, J., Kikuchi, T., Honda, M. and Tachibana, Y. (2011). Recurrence of intraseasonal cold air outbreak during the 2009/2010 winter in Japan and its ties to the atmospheric condition over the Barents-Kara Sea. *Science Online Letters on the Atmosphere*, 7, 25–28.

Iijima, Y. and Hori, M. E. (2017). Cold air formation and advection over Eurasia during "dzud" cold disaster winters in Mongolia. *Natural Hazards*, Special Issue "Multiple climate hazards in Eurasian drylands", in press.

Inoue, J., Hori, M. E. and Takaya, K. (2012). The role of Barents Sea ice in the wintertime cyclone track and emergence of a warm-Arctic cold-Siberian anomaly. *Journal of Climate*, 25, 2561–2568.

Institute of Geo-ecology. (2013). *Desertification Atlas*. The Mongolian Academy of Sciences, Ulaanbaatar. 108.

IPCC (Intergovernmental Panel on Climate Change). (2007). *Climate Change 2007: The Physical Science Basis*. Cambridge University Press, Cambridge.

IPCC (Intergovernmental Panel on Climate Change). (2013). *Climate Change 2013: The Physical Science Basis*. Cambridge University Press, Cambridge.

Ishizuka, M., Mikami, M., Yamada, Y., Kimura, R., Kurosaki, Y., Jugder, D., . . . Shinoda, M. (2012). Does ground surface soil aggregation affect transition of the wind speed threshold for saltation and dust emission? *Science Online Letters on the Atmosphere*, 8, 129–132.

Ito, T. Y., Miura, N., Lhagvasuren, B., Enkhbileg, B., Takatsuki, S., Tsunekawa, A. and Jiang, Z. (2006). Satellite tracking of Mongolian gazelles (*Procapra gutturosa*) and habitat shift in their seasonal ranges. *Journal of Zoology*, 269, 291–298.

Ito, T. Y., Tsuge, M., Lhagvasuren, B., Buuveibaatar, B., Chimeddorj, B., Takatsuki, S., . . . Shinoda, M. (2013). Effects of interannual variations in environmental conditions on seasonal range selection by Mongolian gazelles. *Journal of Arid Environments*, 91, 61–68.

Jugder, D. and Shinoda, M. (2011). Intensity of a dust storm in Mongolia during 29–31 March 2007. *Science Online Letters on the Atmosphere*, 7A, 29–31.

Jugder, D., Shinoda, M., Kimura, R., Batbold, A. and Amarjargal, D. (2014). Quantitative analysis on windblown dust concentrations of PM10 (PM2.5) during dust events in Mongolia. *Aeolian Research*, 15, 169–176.

Jugder, D., Shinoda, M., Sugimoto, N., Matsui, I., Nishikawa, M., Park, S.-U., . . . Park, M.-S. (2011). Spatial and temporal variations of dust concentrations in the Gobi Desert of Mongolia. *Global and Planetary Change*, 78(1–2), 14–22.

Jugder, D., Sugimoto, N., Shinoda, M., Kimura, R., Matsui, I. and Nishikawa, M. (2012). Dust, biomass burning smoke, and anthropogenic aerosol detected by polarization-sensitive Mie lidar measurements in Mongolia. *Atmospheric Environments*, 54, 231–241.

Kaczensky, P., Ito, T.Y. and Walzer, C. (2010). Satellite telemetry of large mammals in Mongolia: What expectations should we have for collar function? *Wildlife Biology in Practice*, 6, 108–126.

Kappes, M. (2011). Multi-Hazard Risk Analyses: A Concept and Its Implementation. Ph.D. thesis, University of Vienna, Vienna.

Kappes, M.S., Keiler, M., Von Elverfeldt, K. and Glade, T. (2012). Challenges of analyzing multi-hazard risk: A review. *Natural Hazards*, 64(2), 1925–1958.

Karl, T.R. (1986). The sensitivity of the Palmer drought severity index and Palmer's Z-index to their calibration coefficients including potential evapotranspiration. *Journal of Climate and Applied Meteorology*, 25, 77–86.

Kelly, R.H., Parton, W.J., Hartman, M.D., Stretch, L.K., Schimel, D.S. and Ojima, D.S. (2000). Intra and interannual variability of ecosystem processes in shortgrass steppe. *Journal of Geophysical Research: Atmospheres*, 105(20), 20093–20100.

Keyantash, J. and Dracup, J. (2002). The quantification of drought: An evaluation of drought indices. *Bulletin of the American Meteorological Society*, 83, 1167–1180.

Kimura, R. (2007). Estimation of moisture availability over the Liudaogou river basin of the Loess Plateau using new indices with surface temperature. *Journal of Arid Environments*, 70, 237–252.

Kimura, R. (2012a). Factors contributing to dust storms in source regions producing the yellow-sand phenomena observed in Japan from 1993 to 2002. *Journal of Arid Environments*, 80, 40–44.

Kimura, R. (2012b). Effect of the strong wind and land cover in dust source regions on the Asian dust event over Japan from 2000 to 2011. *Science Online Letters on the Atmosphere*, 8, 77–80.

Kimura, R. (2017). Satellite-based mapping of dust erodibility in northeast Asia. *Natural Hazards*, Special Issue "Multiple climate hazards in Eurasian drylands", in press.

Kimura, R., Bai, L., Fan, J., Takayama, N. and Hinokidani, O. (2007). Evapo-transpiration estimation over the river basin of the Loess Plateau of China based on remote sensing. *Journal of Arid Environments*, 68, 53–65.

Kimura, R., Bai, L. and Wang, J. (2009). Relationships among dust outbreaks, vegetation cover, and surface soil water content on the Loess Plateau of China, 1999–2000. *CATENA*, 77, 292–296.

Kimura, R. and Moriyama, M. (2014). Application of a satellite-based aridity index in dust source regions of northeast Asia. *Journal of Arid Environments*, 109, 31–38.

Kimura, R. and Shinoda, M. (2010). Spatial distribution of threshold wind speeds for dust outbreaks in northeast Asia. *Geomorphology*, 114, 319–325.

Kobayashi, Y., Shimada, A., Nemoto, M., Morita, T., Adilbish, A. and Bayasgalan, M. (2014). Adverse effects of inhaled sand dust particles on the respiratory organs of sheep and goats exposed to severe sand storms in Mongolia. *Folia Histochemica Cytobiologica*, 52(3), 244–249.

Koike, T. and Shinoda, M. (2011). Temperature changes before and after precipitation in Mongolia. *Journal of Agricultural Meteorology*, 67(3), 139–150.

Koike, T., Shinoda, M. and Morinaga, Y. (2014). Weather conditions of cold seasons related to the recent high mortality of livestock at Bulgan in Northern Mongolia. *Climate in Biosphere*, 14, 29–40 (In Japanese with English abstract).

Koike, T., Shinoda, M., Morinaga, Y. and Gomboluudev, P. (2010). Cold- and warm-deep-snow winters in Mongolia. *Journal of Agricultural Meteorology*, 66(2), 103–110.

Komiyama, H. (2005). Actual damage to Mongolian animal husbandry due to the 2000–2002 *dzud* disaster. *Bulletin of Japanese Association for Mongolian Studies*, 35, 73–85 (In Japanese).

Komiyama, H. (2013). Looking back on the 2010 *dzud* (cold and snow disaster) in Mongolia. *Nihon to Mongol*, 126, 33–38 (In Japanese).

Kurosaki, Y. and Mikami, M. (2005). Regional difference in the characteristic of dust event in East Asia: Relationship among dust outbreak, surface wind, and land surface condition. *Journal of the Meteorological Society of Japan*, 83A, 1–18.

Kurosaki, Y., Shinoda, M., Mikami, M. and Nandintsetseg, B. (2011a). Effects of soil and land surface conditions in summer on dust outbreaks in the following spring in a Mongolian grassland. *Science Online Letters on the Atmosphere*, 7, 69–72.

Kurosaki, Y., Shinoda, M. and Mikami, M. (2011b). What caused a recent increase in dust outbreaks over East Asia? *Geophysical Research Letter*, 38, L11702.

Kurushin, R.A., Bayasgalan, A., Olziybat, M., Enhtuvshin, B., Molnar, P., Bayarsayhan, Ch., . . . Lin, J. (1997). The surface rupture of the 1957 Gobi-Altay, Mongolia, earthquake. *Geological Society of America Special Paper*, 320, 143.

Lancaster, N. and Baas, A. (1998). Influence of vegetation cover on sand transport by wind: Field studies at Owens Lake, California. *Earth Surface Processes and Landforms*, 23, 69–82.

Lepers, E., Lambin, E. F., Janetos, A. C., DeFries, R., Achard, F., Ramankutty, N. and Scholes, R. J. (2005). A synthesis of rapid land-cover change information for the 1981–2000 period. *Bioscience*, 55(2), 115–124.

Levine, M. (1999). The origin of horse husbandry on the Eurasian Steppe. In Levine, M., Rassamakin, Y., Kislenko, A. and Tatarintseva, N. (eds.), *Late Prehistoric Exploitation of the Eurasian Steppe*. McDonald Institute for Archeological Research, Cambridge. 5–58.

Liang, T. G., Liu, X. Y., Wu, C. X., Guo, Z. G. and Huang, X. D. (2007). An evaluation approach for snow disasters in the pastoral areas of Northern Xinjiang, PR China. *New Zealand Journal of Agricultural Research*, 50, 369–380.

Lotsch, A., Friedl, M.A., Anderson, B. T. and Tucker, C. J. (2005). Response of terrestrial ecosystems to recent Northern Hemispheric drought. *Geophysical Research Letters*, 32, L06705.

MacLachlan, I. (2006). The Historical Development of Cattle Production in Canada. https://www.uleth.ca/dspace/handle/10133/303 (Accessed: 20 December 2014).

Mandakh, N., Dash, D. and Khaulenbek, A. (2007). Present status of desertification in Mongolia. In Tsogtbaatar, J. (ed.), *Geoecological Issues in Mongolia*. Institute of Geoecology of Mongolian Academy of Sciences, Ulaanbaater. 63–73.

Matsushima, D., Kimura, R. and Shinoda, M. (2012a). A method for estimating threshold wind speed of dust emission using thermal inertia over a non-vegetated surface. *Journal of Japan Society of Civil Engineers, Ser. B1*, 68(4), 1789–1794 (in Japanese).

Matsushima, D., Kimura, R. and Shinoda, M. (2012b). Soil moisture estimation using thermal inertia: Potential, practical feasibility, and sensitivity to data conditions. *Journal of Hydrometeorology*, 13(2), 638–648.

McKee, T., Doeskan, N. and Kleist, J. (1993). The Relationship of Drought Frequency and Duration to Time Scales. *Eighth conference on applied climatology*. American Meteorological Association, Anaheim, CA, 179–184.

Middleton, N., Rueff, H., Sternberg, T., Batbuyan, B. and Thomas, D. (2015). Explaining spatial variations in climate hazard impacts in Western Mongolia. *Landscape Ecology*, 30(1), 91–107.

Miyazaki, S., Yasunari, T., Miyamoto, T., Kaihotsu, I., Davaa, G., Oyunbaatar, D., . . . Oki, T. (2004). Agrometeorological conditions of grassland vegetation in central Mongolia and their impact for leaf area growth. *Journal of Geophysical Research*, 109, D22106.

Mohamed, A.A., Kimura, R., Shinoda, M. and Moriyama, M. (2011). Diurnal surface temperature difference index derived from ground-based meteorological measurements for assessment of moisture availability. *Journal of Arid Environments*, 75, 156–163.

Morinaga, Y., Bayarbaator, L., Erdenetsetseg, D. and Shinoda, M. (2004). Zoo-Meteorological Study of Cow Weight Change in Forest Steppe Region, Mongolia. *The sixth international workshop on climate change in arid and semi-arid region of Asia*. Ulaanbaatar, Mongolia. 82–83.

Morinaga, Y., Tian, S. and Shinoda, M. (2003). Winter snow anomaly and atmospheric circulation in Mongolia. *International Journal of Climatology*, 23, 1627–1636.

Mu, H., Battsetseg, B., Ito, T., Otani, S., Onishi, K. and Kurozawa, Y. (2010). Effects of Asian dust storm on health-related quality of life: A survey immediately after an Asian dust storm event in Mongolia. *International Journal of Health Research*, 3, 87–92.

Mu, H., Battsetseg, B., Ito, T., Otani, S., Onishi, K. and Kurozawa, Y. (2011). Health effect of dust storms: Subjective symptoms of eyes and respiratory system in inhabitants in Mongolia. *Journal of Environmental Health*, 73, 18–20.

Mu, H., Otani, S., Shinoda, M., Yokoyama, Y., Onishi, K., Hosoda, T., . . . Kurozawa, Y. (2013). Long-term effect of livestock loss caused by dust storms on Mongolian inhabitants: A survey 1 year after the dust storm. *Yonaga Acta Medica*, 56, 39–42.

Munich Re (2011). NATHAN (Natural Hazards Assessment Network) World Map of Natural Hazards. https://www.munichre.com/site/corporate/get/documents/mr/assetpool.shared/Documents/0Corporate%20Website/Publications/302–05972en.pdf (Accessed: 20 December 2014).

Nakano, T., Nemoto, M. and Shinoda, M. (2008). Environmental controls on photosynthesis production and ecosystem respiration in semi-arid grasslands of Mongolia. *Agricultural and Forest Meteorology*, 148, 1456–1466.

Nakano, T. and Shinoda, M. (2010). Response of ecosystem respiration to soil water and plant biomass in a semi-arid grassland. *Soil Science and Plant Nutrition*, 56(5), 773–781.

Nandintsetseg, B. and Shinoda, M. (2011). Seasonal change of soil moisture and its climatology and modeling in Mongolia. *International Journal of Climatology*, 31, 1143–1152.

Nandintsetseg, B. and Shinoda, M. (2013). Assessment of drought frequency, duration, and severity and its impact on pasture production in Mongolia. *Natural Hazards*, 66, 995–1008.

Nandintsetseg, B. and Shinoda, M. (2014). Multi-decadal soil moisture trends in Mongolia and their relationships to precipitation and evapotranspiration. *Arid Land Research and Management*, 28, 247–260.

Nandintsetseg, B. and Shinoda, M. (2015). Land surface memory effects on dust emission in a Mongolian temperate grassland. *Journal of Geophysical Research: Biogeosciences*, 120 (3), 414–427.

Nandintsetseg, B., Shinoda, M., Kimura, R. and Ibaraki, Y. (2010). Relationship between soil moisture and vegetation activity in the Mongolian Steppe. *Science Online Letters on the Atmosphere*, 6, 29–32.

Nandintsetseg, B., Shinoda, M. and Shao, Y. (2014). Land Surface Memory Effects on Dust Emission and Its Modeling in a Mongolian Temperate Grassland. *JSPS 1st meeting on outbreaks of Asian dust and environmental regime shift*. Nagoya University, Nagoya.

Natsagdorj, L. (2003). Climate change. In Batima, P. (ed.), *Climate Change: Pasture and Animal Husbandry*. Institute of Meteorology and Hydrology of Mongolia, Ulaanbaatar. 13–14.

Natsagdorj, L. and Dulamsuren, J. (2001). Some aspects of assessment of the *dzud* phenomena. *Papers in Meteorology and Hydrology*, 23, 3–18.

Obasi, G.O.P. (1994). WMO's role in the international decade for natural disaster reduction. *Bulletin of the American Meteorology Society*, 75, 1655–1661.

Otani, S., Mu, H., Onishi, K., Kurozawa, Y. and Shinoda, M. (2014). Risk Prediction of the Effects of Severe Winter Disasters (*Dzud*) on Nomadic Health and Society in Mongolia with Regard to Loss of Livestock. https://wce.confex.com/wce/2014/webprogram/Paper2334.html (Accessed: 7 October 2014).

Ozaki, T. (2010). Analysis on a report of natural disaster in Mongolia–A case study of storm (*Shuurga*) in May 2008. *Historical Studies of Kagoshima University*, 57, 9–23.

Palmer, W. C. (1965). Meteorological drought. *US Weather Bureau Research Paper*, 45, 58.

Pelling, M. (2014). Review of global risk index projects: Conclusions for sub-national and local approaches. In Birkmann, J., Kienberger, S. and Alexander, D. (eds.), *Assessment of Vulnerability to Natural Hazards: A European Perspective*. Elsevier, San Diego, CA. 151–170.

Pfisterer, A. B. and Schmid, B. (2002). Diversity-dependent production can decrease the stability of ecosystem functioning. *Nature*, 416, 84–86.

Putkonen, J. and Roe, G. (2003). Rain-on-snow events impact soil temperatures and affect ungulate survival. *Geophysical Research Letters*, 30(4), 1188.

Robinson, S. and Milner-Gulland, E. J. (2003). Political change and factors limiting numbers of wild and domestic ungulates in Kazakhstan. *Human Ecology*, 31, 87–110.

Saltelli, A., Chan, K. and Scott, E. M. (2000). *Sensitivity Analysis*. John Wiley & Sons, New York.

Schneider, S. H. and Dickinson, R. E. (1974). Climate modeling. *Reviews of Geophysics and Space Physics*, 12(3), 447–493.

Schneiderbauer, S. and Enrlich, D. (2004). *Risk, Hazard and People's Vulnerability to Natural Hazards: A Review of Definitions, Concepts and Data*. Office for Official Publication of the European Communities, Luxembourg.

Shang, Z. H., Gibb, M. J. and Long, R. J. (2012). Effect of snow disasters on livestock farming in some rangeland regions of China and mitigation strategies – A review. *Rangeland Journal*, 34, 89–101.

Shao, Y. and Dong, C. H. (2006). A review on East Asian dust storm climate, modeling and monitoring. *Global Planetary Change*, 52, 1–22.

Shinoda, M. (2001). Climate memory of snow mass as soil moisture over central Eurasia. *Journal of Geophysical Research*, 106, 33393–33403.

Shinoda, M. (2012). Land: Proactive management of drought and its derived disasters. In Shaw, R. and Phong, T. (eds.), *Environment Disaster Linkages: Community, Environment and Disaster Risk Management*, 9. Emerald Publishers, Bingley. 61–78.

Shinoda, M. (2015). High-Impact Weathers in a Changing Climate Over Arid Eurasia and Proactive Disaster Management. Procedia IUTAM, *ScienceDirect* 17. 47–52. http://www.sciencedirect.com/science/article/pii/S2210983815002461

Shinoda, M., Gillies, J. A., Mikami, M. and Shao, Y. (2011). Temperate grasslands as a dust source: Knowledge, uncertainties, and challenges. *Aeolian Research*, 3, 271–293.

Shinoda, M., Ito, S., Nachinshonhor, G. U. and Erdenetsetseg, D. (2007). Phenology of Mongolian grasslands and moisture conditions. *Journal of Meteorological Society of Japan*, 85(3), 359–367.

Shinoda, M., Kimura, R., Mikami, M., Tsubo, M., Nishihara, E., Ishizuka, M., . . . Kurosaki, Y. (2010a). Characteristics of dust emission on the Mongolian Steppe: The 2008 DUVEX intensive observational period. *Science Online Letters on the Atmosphere*, 6, 9–12.

Shinoda, M. and Morinaga, Y. (2005). Developing a combined drought-*dzud* early warning system in Mongolia. *Geographical Review of Japan*, 78, 928–950 (In Japanese with English abstract).

Shinoda, M., Nachinshonhor, G. U. and Nemoto, M. (2010b). Impact of drought on vegetation dynamics of the Mongolian Steppe: A field experiment. *Journal of Arid Environments*, 74, 63–69.

Shinoda, M. and Nandintsetseg, B. (2011). Moisture and vegetation memories in a cold, arid climate. *Global Planetary Change*, 79, 110–117.

Shinoda, M., Nandintsetseg, B., Nachinshonhor, U. G. and Komiyama, H. (2014). Hotspots of recent drought in Asian Steppes. *Regional Environmental Change*, 14(1), 103–117.

Shinoda, M. and Yamaguchi, Y. (2003). Influence of soil moisture anomaly on temperature in the Sahel: A comparison between wet and dry decades. *Journal of Hydrometeorology*, 4(2), 437–447.

Sternberg, T. (2010). Unravelling Mongolia's extreme winter disaster of 2010. *Nomadic Peoples*, 14, 72–86.

Sternberg, T., Middleton, N. and Thomas, D. (2009). Pressurized pastoralism in South Gobi Province, Mongolia: What is the role of drought? *Transactions of British Geographers – IBG*, 34, 364–377.

Sternberg, T., Thomas, D. and Middleton, N. (2011). Drought dynamics on the Mongolian Steppe 1970–2006. *International Journal of Climatology*, 31, 1823–1830.

Tachiiri, K. and Shinoda, M. (2012a). Impact of future climate change on summer droughts and snow disasters in Mongolia. *Science Online Letters on the Atmosphere*, 8, 124–128.

Tachiiri, K. and Shinoda, M. (2012b). Quantitative risk assessment for future meteorological disasters: Reduced livestock mortality in Mongolia. *Climatic Change*, 113, 867–882.

Tachiiri, K. and Shinoda, M. (2014). Developing a livestock model for drought and *dzud* risk assessment in Mongolia. *International symposium on multi-hazard approach in Mongolia*. Nagoya University, Nagoya.

Tachiiri, K., Shinoda, M., Klinkenberg, B. and Morinaga, Y. (2008). Assessing Mongolian snow disaster risk using livestock and satellite data. *Journal of Arid Environment*, 72, 2251–2263.

Takeda, S., Tanaka, H., Shimada, A., Morita, T., Ishihara, A., Adilbish, A., . . . Gungaa, O. (2014). Cerebellar ataxia suspected to be caused by *oxytropis glabra* poisoning in Western Mongolian goats. *The Journal of Veterinary Medical Science*, 76, 839–846.

Tilman, D. and Downing, J.A. (1994). Biodiversity and stability in grasslands. *Nature*, 367, 363–365.

Tsedevsuren, D., Mijiddorj, R., Namkhai, A. and Enhbat, D. (1997). *Catalogue of Droughts and Dzud in Central and Eastern Part of Mongolia*. Hydrometearological Service Printing, Ulaanbaatar, 105.

Turner II, B. L., Kasperson, R. E., Matson, P. A., McCarthy, J. J., Correll, R. W., Christensen, L., . . . Schiller, A. (2003). A framework for vulnerability analysis in sustainability science. *Proceedings of the National Academy of Sciences*, 100, 8074–8079.

UN. (2002). Johannesburg Plan of Implementation of the World Summit on Sustainable Development. *Tech. rep.*, United Nations. http://www.un.org/esa/sustdev/documents/WSSDPOIPD/English/WSSDPlanImpl.pdf (Accessed: 3 September 2009).

UNDP (Government of Mongolia and United Nations Development Programme). (2008). Meteorological Disaster Event, Ulaanbaatar. Mongolia (In Mongolian).

UNEP. (1992a). Agenda 21. United Nations Environment Programme. http:// www.un.org/esa/dsd/agenda21/res_agenda21_07.shtml (Accessed: 3 September 2009).

UNEP. (1992b). *World Atlas of Desertification*. Arnold, London.

UNEP. (1992c). Desertification, Land Degradation [Definitions]. Desertification Control Bulletin 21, UNEP, Nairobi.

UNEP. (2004). *Reducing Disaster Risk: A Challenge for Development: A Global Report.* UNEP-BRCP, New York.

UN-ISDR. (2005). Hyogo Framework for Action 2005–2015: Building the Resilience of Nations and Communities to Disasters. *World conference on disaster reduction.* Hyogo, Japan.

Vinnikov, K.Y., Robock, A., Speranskaya, N.A. and Schlosser, C.A. (1996). Scales of temporal and spatial variability of midlatitude soil moisture. *Journal of Geophysical Research*, 101, 7163–7174.

Wang, W., Liang, T., Huang, X., Feng, Q., Xie, H., Liu, X., . . . Wang, X. (2013). Early warning of snow-caused disasters in pastoral areas on the Tibetan Plateau. *Natural Hazards Earth System Science*, 13, 1411–1425.

Wilhite, D.A. (2000). Drought as a natural hazard: Concepts and definitions. In Wilhite, D. A. (ed.), *Drought, a Global Assessment*. Routledge, London. 3–18.

Wisner, B., Blaikie, P., Cannon, T. and Davis, I. (2004). *At Risk: Natural Hazards, People's Vulnerability and Disasters*. 2nd ed. Routledge, London.

WMO. (1999). Comprehensive Risk Assessment for Natural Hazards. *Technical document 955*. World Meteorological Organisation. http://www.planat.ch/ressources/planat producten198.pdf (Accessed: 5 May 2010).

Zhao, M.L., Gao, X. and Han, B. (2011). Locoweed poisoning in the native grassland of China. *International Journal of Poisonous Plant Research*, 1, 41–46.

6 Climate change and security

Major challenges for Yemen's future[1]

Helen Lackner

Introduction

In 2015 Yemen sank into civil war, and its state structures effectively collapsed, thus raising the question as to the country's ability to address the major environmental problems, which threaten its very existence in the long term. This followed years of seemingly endless major political, economic and social crises without any likely solutions. Regardless of the outcome of the war, which is worsened by foreign involvement, climate change is a fundamental challenge which will constrain the country's national development in coming decades regardless of solutions to the immediate political crises. Water scarcity, in particular, is already making international headlines, without resulting in an adequate response from decision makers. Although Yemeni politicians have limited influence over climate change, they could do much more to mitigate its impact. While they fight over short-term political rivalries at the expense of the suffering population, the country's very survival is under threat due to the lack of serious attention to its major environmental challenges, water shortage in particular.

This chapter examines the synergy and connections between climate change, environmental deterioration, worsening socio-economic indicators and the pre-war aspects of the equally poor security situation. In conclusion, the chapter will make suggestions of activities and approaches which could contribute to improvements and ensure that Yemenis have a better life in coming decades.

Security aspects will be discussed not only with respect to the threat posed by international Islamist insurgent groups, but also with respect to the daily security problems faced by ordinary Yemenis. Changes brought about by the full-scale war affecting many parts of the country will be touched upon but not analysed in detail. Security has many meanings. In this chapter, I will address three issues:

- food security and the ability of families to survive, satisfy their basic needs and achieve an acceptable living standard.
- security and safety for people to go about their daily business without risking life and limb
- problems brought about by the various armed conflicts, which have affected Yemen in the past two decades.

Impact of climate change on the daily life of Yemenis, particularly rural Yemenis, will be described to understand better the extent to which the country's changing natural environment, increasingly unpredictable weather and social transformation are reducing people's security. The chapter will also address the state's response to these issues and the impact and influence of neo-liberal development policies which have reduced the role of the state and the provision of basic services to the population.

Background[2]

Situated at the southwestern corner of the Arabian Peninsula, Yemen used to be known as 'Arabia Felix' thanks to the monsoon rains which watered its western highlands and provided a sharp contrast to the aridity of the rest of the Peninsula (Figure 6.1). Since unification in 1990, when the Republic of Yemen was established, the country's population has more than doubled from 11.3 million[3] to an estimated 26 million in 2015.[4] By the end of 2015, over 21 million people were in need of some kind of assistance, 20 million did not have access to adequate water or sanitation, and 13 million, or 50 per cent of the population, were food insecure,[5] 15 million lacked access to health care and more than 1 million children were malnourished – including 15 per cent of children under five suffering from Global Acute Malnutrition (GAM).[6]

Between the summer of 2013 and February 2014, Saudi Arabia deported over half a million Yemenis lacking the correct documentation for residence and work, further exacerbating conditions in Yemen. Though, in 2015, Saudi Arabia allowed thousands of Yemenis in and gave them work permits, thus ensuring that their

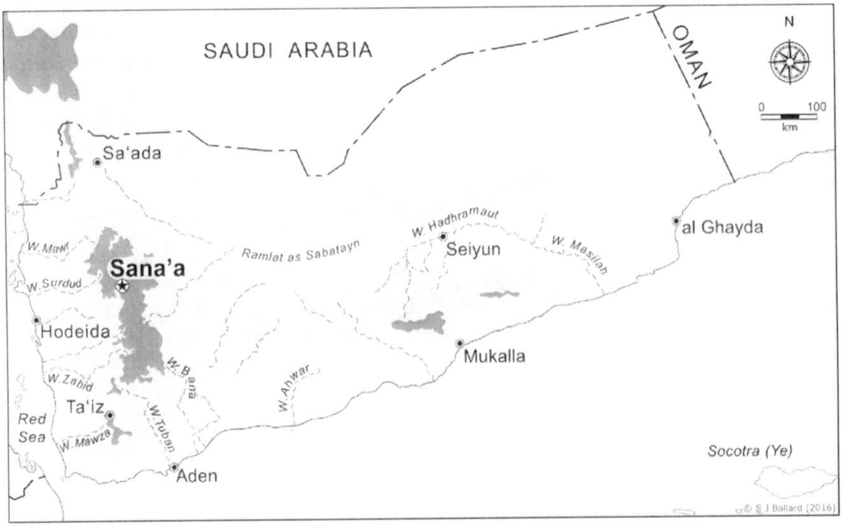

Figure 6.1 Map of Yemen

Source: Sebastian Ballard (2016)

presence and activities could be monitored. The various internal conflicts displaced over half a million people between 2011 and 2014 and, in 2015 prior to the war, about 300,000 people remained displaced, the vast majority of them in the far north of the country. Since then there have been considerable population movements, mainly from the cities to rural areas and away from the areas of heaviest fighting.

In addition, the country hosts over 1 million refugees,[7] mostly from the Horn of Africa and other Middle Eastern countries (Palestine, Iraq, Syria). Poverty is officially estimated to affect about 55 per cent of the population and over 80 per cent of the poor are in rural areas, which are still home of almost 70 per cent of the total population. Over half the population are under 18 years of age and 43 per cent under 15; so although only 3 per cent of the population are over 65, the dependency ratio is very high. While educational facilities have increased considerably in the past two decades, illiteracy is still widespread with an adult male literacy rate of 83 per cent but only 52 per cent for women in 2013.[8] The low quality of education is a further problem with only a small percentage of primary school graduates enrolling in secondary school and the standards of university and further education significantly below those needed for economic expansion. Health indicators are little better, though infant mortality rates have dropped significantly to about 46 per 1,000 births. At the best of times Yemen's rulers face a very difficult situation.

These are not the best of times. After having long been largely neglected, Yemen has recently come into media prominence first as a result of the full-scale war involving Saudi Arabia and its coalition of international partners in support of the transitional government elected in 2012 following the 2011 popular uprisings, fighting against the former president allied with insurgents originally from the far north of the country. In addition to the war, US drones[9] have continued to target Islamist insurgents and also constantly fly over rural areas spreading fear and occasional destruction. Numerous factions are involved in different parts of the country.

It is in this broad context that the country's population have to address the additional challenges of climate change, largely without the support of an effective state. Although Yemen is only a minor contributor to climate change due to its low use of energy[10] and its relatively small population, as is the case elsewhere, its people are likely to suffer comparatively more since most of them are poor and are primarily dependent on agriculture and primary sector economic production. As

> global warming is likely to increase the intensity of droughts, floods, hurricanes, storms and wildfires . . . poor people will suffer most, as famines become more frequent and marginal lands become even less viable. . . . Climate change . . . is highly likely to have a severely negative impact on societies that already suffer the greatest risks to their water supplies.[11]

Yemen is a perfect example of this situation.

The most important and fundamental of these is the water crisis that is clearly environmentally related, and some of its features are recognised as being the effect of climate change; it is also one which has immediate and direct security implications insofar as smaller and larger conflicts over water are daily occurrences. In

the wake of the 2011 uprisings and as the political crisis shows no sign of abating, political discourse is entirely focused on the distribution of power between the factions involved and gives far too little attention to the major issues of development and livelihoods of the majority of Yemen's population. This is despite the fact that poverty has become endemic and people's living conditions are deteriorating further due to climate-related agro-ecological changes, worsening insecurity and economic collapse. Until recently, neither government nor foreign development institutions gave much attention to the major environmental issues, even the best known one, water. It is only since the turn of the century that the importance and dangers presented by these factors have begun to enter the national political discourse.

Nor has the broadly advertised policy dialogue between Yemen and its development partners addressed this fundamental synergy. Until the early 2000s, funding agencies supported development policies focused on short-term objectives that fit neatly into the Washington Consensus philosophy. To a considerable extent, these policies gave priority to investments and activities that contributed to reducing the long-term sustainability of Yemen's ecology and its ability to support its population.[12] In agriculture, for example, research focused on irrigation and export crop development. While there are now attempts to address the socio-economic implications of climate change, this chapter examines their significance, the extent to which they represent a real change in government and aid agency approaches and the likelihood of their having a meaningful impact on the fundamental long-term crisis that threatens the very survival of the country. Impact of the war is not discussed in detail but is likely to worsen many of the situations described below.

Environmentally related challenges

With greatly underdeveloped industrial and tertiary sectors, Yemen is particularly vulnerable to climate hazards as 70 per cent of its population are rural and over 55 per cent are dependent on agriculture and livestock husbandry, as well as a further 5 per cent or so on fisheries. While climate change is a recognised fact, its impact cannot be predicted with any certainty. Three main possible scenarios have been envisaged, with the most likely being 'hot and dry' and 'warm and wet.' A recent World Bank study[13] predicts that regardless of which scenario actually happens, 'Yemen will be getting warmer, most likely at a faster rate than the global average . . . there will be more variability of rainfall patterns within years [and] there will probably be an increased frequency of intense rainfall events and therefore possibly an increased risk of floods.'[14]

Water

Water is the fundamental issue affecting the future of Yemen.[15] In my view[16] it is the most important problem Yemenis must address if future generations are to continue living in the country. The shortage of water in Yemen is absolute: currently annual use of water at 3.5 billion m^3 exceeds annual renewable resources by

1.4 billion m^3, in other words, one third of the water used annually is mined from a fossil aquifer. Annual per capita availability of renewable water has now dropped to less than $85m^3$, which is significantly below 10 per cent of what is internationally accepted as the scarcity threshold ($1000m^3$). The World Bank estimates that groundwater reserves are likely to be depleted in about three decades.[17]

Yemen is almost exclusively dependent on groundwater. The renewable aquifers depend on annual and also more long-term rainfall. Climate change is an important factor in the depletion of the water source: with more violent and more irregular rainfall episodes, the amount of the water which replenishes the various shallow aquifers is reduced as the rush of the flow gives insufficient time for the water to penetrate the soil and be absorbed. Hence, the replenishment to rainfall ratio is going down just at a time when the need and demand for water are increasing as a result of population increase, improved hygiene requirements and many other relevant factors. Other causes are social and economic, such as the use of powerful pumps in agriculture. While water issues in Yemen deserve (and are partly getting) books onto themselves,[18] here we will move on to the other climate-related factors affecting people's living conditions and, therefore, security.

Lack of clean drinking water is a well-known, serious health hazard and has an influence on productivity and well-being.[19] Not only are many urban areas not covered by public water distribution systems, but also the tanker loads purchased from private suppliers are polluted. In rural areas, the cost of water is prohibitive for the poor who have to buy it, and extremely time consuming for women and children who collect it from wells, irrigation pumps, springs and standpipes. Most of this water is also not potable and presents serious health risks.

With the increasing scarcity of water in the highlands, population movements towards the coastal areas have already started taking place on a temporary as well as a more permanent basis; this shift is likely to increase in coming decades. When wells that supply domestic water dry up, households are forced either to purchase water from tankers which travel increasingly long distances and thus water becomes more and more expensive, or they have to move. Many families move temporarily to stay with relatives in villages or towns where water is still available. They then return home after good rains when the wells have filled. There are already cases of villages being permanently abandoned in some highland governorates such as Amran, al Baidha, Dhala' and Sa'ada.[20] The major cities of the lowlands are likely to see significantly increased populations as a result of such movements. For example, many al Baidha families have close relations in Hodeida, and are therefore likely to head in that direction.

Surprisingly, desalination has barely featured in Yemeni planning to date. The National Water Sector Strategy and Investment Programme[21] (NWSSIP) of 2005 barely mentions desalination, with a total proposed investment of as little as US$500,000 for studies of desalination, and solar and wind energy! Since then there has been a proposed public–private project to provide desalinated water to Taiz, the city which has had the worst water crisis for well over two decades, but this has not proceeded beyond planning.[22] This is despite the fact that desalination of water for domestic purposes is clearly essential if people are to continue living

in Yemen later this century. While the cost of desalination is dropping thanks to improved technologies, it remains a major issue for Yemen which does not have the financial means of its neighbours in the Peninsula.

Saline intrusion into the coastal aquifers used both for human consumption and for agriculture is another problem. Three of Yemen's major cities (Hodeida, Aden and Mukalla) are on the coast and their populations are increasing rapidly as a result of rural people being forced out of the highlands when water supplies are completely exhausted. The over-exploitation of the rural aquifers in the hinterland of these cities is already causing serious problems to life and agriculture in those areas in addition to the fact that these supplies are rapidly being exhausted. This is another cause for insecurity through conflict between rural and urban demand. There already is a documented history of rural-urban conflicts over water in connection with Taiz.[23] In addition, the risk of rising sea levels is a further climate change hazard that planners should take into consideration as it will certainly affect an increasingly large proportion of the country's population and will require massive mitigating infrastructure investments. By their very geographical location, the population of the aforementioned coastal cities would be among those most affected, as well as smaller towns and fishing communities and villages.

Weather events

One of the most common indicators/symptoms of climate change has been the changing pattern of rainfall. Yemen is on the edge of the monsoon zone, and its rainy seasons in the highlands (the only areas with regular rainfall) are determined by the monsoons. Whereas the rainy seasons used to be relatively well defined, falling in March–April and July–August,[24] recent decades have seen significant changes with far more unpredictability of both timing and intensity. Given that over 60 per cent of Yemen's agriculture is still rainfed, the unpredictability of rains has a major impact on people's livelihoods as, in many cases, farmers prepare the land and even sow seeds before the rains; their labour and financial input are wasted when the rains fail completely and returns are very limited when the rains are insufficient or fall at the wrong time, thus leading to crops never reaching maturity. Although leaves and stalks are then used as animal fodder, this is not as useful as having grain crops for human consumption whether sorghum, maize, wheat or any other crop.

Violent and sudden rains damage crops as well as washing away soil, thus causing short-term suffering to people and long-term deterioration of the potential of their land. Violent downpours reduce the amount of water absorbed by the soil and available to replenish the water table. The year 2015 will be remembered in Yemen not only for the start of the war but also as a unique one, climatically. In addition to all its other problems, the country suffered not one, but two unprecedented and extremely violent cyclones in November: Chapala and Megh. Chapala brought hurricane force winds of over 120km/h, 610mm of rain in 48 hours, seven times more than the annual average, and displaced some 45,000 people, causing massive destruction mostly on the ecologically unique Socotra island and in Hadramaut and Shabwa governorates. After causing further destruction in Socotra, Megh

weakened but went to hit more western parts and even reached the highlands in Ibb governorate. These may well be further indicators of climate change in the region.

Data on weather events are not easily accessible. The following table gives an indication of major floods in the past two decades.

The graph in Figure 6.2 indicates the frequency of droughts in the past century, clearly showing a worsening pattern since 1970. Data for more recent

Table 6.1 Major floods 1990–2014

Year	*Event*	*Location/governorate*
1991	Flood	Socotra
1993	Flood	Lahej, Abyan, Aden
1996	Flood	Taiz, Hodeida, Shabwa, Mareb, Hadramaut, Abyan, Jawf
1998	Flood	Tihama/Hodeida, Taizz
1999	Flood	Socotra
2001	Flood	Sa'da, Amran, Hodeida, Hadramaut,
2002	Flood	Tihama, Taiz, Hadramaut, Rayma
2003	Flood	Hajja, Taiz
2005	Flood	Sana'a, Hodeida
2006	Flood	Dhamar, Hodeida, Sana'a, Taizz, Sa'da
2007	Flood	Hadramaut, Ibb, Raymah, Dhamar
2008	Flood	Hadramaut, Mahara, Taizz, Lahej, Mahweet,
2010	Flood	Sana'a, Ibb, Dhamar,
2013	Flood	Taizz, Dhamar, Mahweet, Sana'a, Ibb, Hajja, Hodeida, Shabwa, Abyan

Source: EM-DAT (2014)

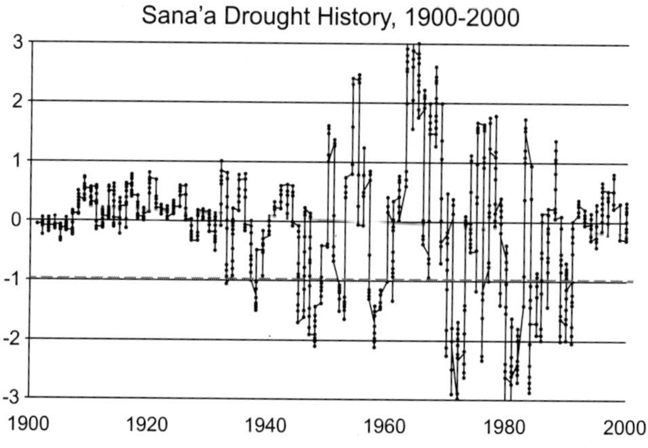

Figure 6.2 Sana'a drought history calculated with the Standard Precipitation Index. Less than -1 (dotted line) signifies drought. Data from the UK Climate Research Unit (Harris et al. 2014)

years are not available, but there have been records of droughts every few years and particularly in 2008 and 2014 in some areas, and almost everywhere in the country in 2009. In Yafi', a major coffee-growing area, there was a drought lasting almost 10 years in the first decade of the century, which killed off the coffee bushes.

While the information is only partial, the frequency of occurrences shows that such events have a significant impact on the country's development. All climate-related emergencies incur heavy costs of different types:

- reconstruction of damaged infrastructure is expensive and uses funds which could otherwise be devoted to new development;
- lost lives are irreplaceable and of course affects relatives left behind whose living conditions are worsened by the loss of labour power of those dead in addition to the emotional suffering involved
- lands washed away reduce production potential and thus negatively affect production, incomes and nutrition
- all these are forms of insecurity that have social and psychological consequences on affected populations.

Soil deterioration and erosion

Only 3 per cent of the country's surface is suitable for agricultural use, including pastureland. Hence soil erosion and desertification are further problems, worsened by climate change. With increasing periods of drought, good topsoil is blown away by the wind. With violent storms, the banks of *wadis* are swept away, *wadis* widened and agricultural land, homes as well as infrastructure destroyed. This also causes floods further downstream when the banks are unable to withstand the violence of flash floods caused by sudden downpours. Rapid flows reduce the replenishment of the water table.

Desertification is a phenomenon, which has increased and particularly affects the areas that are closest to the existing deserts. It is estimated that 3–5 per cent of agricultural land is lost annually in this way.[25] While certain archaeological sites which are currently well into the deserts (Mareb, Shabwa, Jawf) were irrigated agricultural areas many centuries ago, the deserts continue to expand and increasing areas of the governorates bordering the Rub' al Khali desert shift initially from cultivation to rangeland and eventually to completely desertified areas. This concerns the governorates of Jawf, Mareb, Shabwa, Hadramaut and Mahra primarily.

Security issues

The most immediate form of insecurity related to the environment concerns the shortage of water. While no reliable figures are available,[26] any Yemeni can provide a multiplicity of anecdotes about such issues, including armed confrontations. Without water there is no life, and as the shortage of water worsens, people's

desperation to access what is there increases and so does the number and seriousness of conflicts. This is causing a whole range of security issues including:

- access to domestic water from wells or pumps; with drops in the water table, wells dry up or their output reduces significantly. Women and children spend increasingly long periods waiting for water, often in the sun and heat and with the prospect of carrying heavy 20 l jerry cans up hills to their homes or, at best, having to load and later unload, five or six of these on the backs of their donkeys. These waits, in addition to being time largely wasted, are opportunities for the emergence of smaller and larger conflicts as people try to jump the queue or older disagreements are revived. Daily stress caused in this way can occasionally develop into armed confrontations between neighbours.
- Disagreements about payment for domestic water supplies.
- Allocation of spate water for irrigation is another opportunity for small-scale conflicts, as disagreements occur between farmers for taking too much water or diverting someone else's water to their fields. This again is liable to develop into physical violence.
- Disagreements about ownership of springs and wells between individuals and social groups
- The general phenomenon of shallow wells drying out as a result of wealthier people deepening their boreholes is a major change. These incidents are becoming increasingly frequent in some areas, particularly in the northern highlands, as the availability of water is reducing and people are concerned both with water for domestic uses and for crop cultivation. This is yet another symptom of the increasing gap between the few wealthy having the opportunity of becoming more so, while the majority of the poor get poorer, as their one economic asset becomes increasingly worthless, as land without water cannot be cultivated. This produces not only conflicts between the owners of shallow wells and those with deep boreholes but can also involve issues between different owners of boreholes as the cost of deepening becomes increasingly unaffordable.

Food insecurity is closely connected to water insecurity. Without water, people cannot cultivate their land; this can be due either to irregular and unseasonal rainfall or to the aforementioned lack of water resulting from inadequate access to groundwater or spates. Another climate-related aspect, which reduces food security, is soil deterioration, including both the disappearance of soils through wind and water erosion as well as pollution and reduced fertility of soils due to bad management. In Yemen, while pesticides and herbicides play a minor role, the lack of natural fertilizers and bad cropping patterns are also contributors to soil deterioration. Thus people's food production is reduced and their ability to buy food is also reduced when they have no produce for sale, either from crop production or from their livestock. Given that Yemen currently imports over 90 per cent of its basic staples (wheat, rice, sugar, tea), it is clear that most food is purchased.

With the expected continued rise in world food prices, food insecurity is likely to worsen over the coming decades particularly affecting the poor and increasing poverty levels in the country.

Other forms of insecurity also affect people's lives daily: robberies at home, attacks on travellers and other forms of theft. While basic criminality is not directly related to climate change, there is little doubt that worsening poverty and lack of legal income generating potential are underlying reasons which lead some (mostly) men to a criminal life and to stealing either from houses or attacking people on the road. This is in addition to the constant fear people experience about the threat of American drones and the daily worsening encounters between insurgents and official armed and security forces which take place unpredictably in many parts of the country. Since 2015, fear of shelling from one set of factions and aerial bombing from the other are serious contributors to insecurity.

Well-publicised armed encounters between security forces and insurgents have multiplied exponentially since 2011. While Islamist insurgents have been present in the country since the 1990s, the frequency and seriousness of incidents, as well as the numbers of insurgents involved has increased dramatically in the three years before the full-scale war. It is noticeable that the areas where politically motivated insurgents have been most active and involved in the past decade include those where the impact of climate change has been greatest and where poverty rates are very high, and also where sources of income and development investments have been least. Most parts of Sa'ada, Mareb, Jawf, Shabwa and the plateau of Abyan governorates are areas where there have been few development interventions and where insurgents have taken refuge and also where the state has been absent particularly since 2011. The Abyan plateau (Mudia, Mahfadh) and the areas linking it to Shabwa, as well as much of Shabwa itself have been areas where Islamist insurgents have been in their 'comfort zone' for some time. Jawf is a governorate where insecurity has been a feature for many decades. Most of these governorates are experiencing increasing desertification and all are suffering from more irregular rainfall patterns.

The State's response to the double challenge

The policies of the State and its 'international partners' are very closely linked and most government documents have been drafted with considerable involvement from the main external financing agencies, the World Bank in particular. Indeed, the discrepancy between official policy statements very closely influenced by the neo-liberal agenda and the reality implemented on the ground may well, in part at least, be due to the fact that stated policies are not 'owned' by government, i.e. that Yemeni authorities are not committed to the policies they have been pressured into adopting. Further, the government has, on more than one occasion in recent decades, been forced to approve policies which, for good or bad reasons, it does not support: the ongoing saga of fuel subsidies is a prime example of conflicting objectives of the Bretton Woods institutions on the one hand and the State as well as the citizens on the other, though the latter two also have conflicting interests in the matter.

Since the establishment of the Republic of Yemen, policies responding to both climate and security challenges reflect an increasing concern for both. In the 1990s, neither was given much attention: in 1992, the World Bank's only reference to water was to state that 'land and water shortages will continue to severely constrain agriculture.'[27] With the adoption of the Structural Adjustment programme in 1995, half-hearted reforms were implemented, including partial increases in fuel prices (which had, among its explicitly stated aims, reduction of the over-exploitation of groundwater for irrigation), projects to reduce the size of the civil service and the introduction of 'cost-recovery' in health services. All these were publicly advertised as increasing efficiency and reducing corruption.

The first decade of the current century was marked by increased official awareness of environmental issues, primarily of the water crisis, though this did not lead to projects or activities that improved living standards of the poor either in agriculture or in other sectors. These included the NWSSIP for water, the 2009 National Adaptation Programme of Action (NAPA) for the environment which identified seven sectors particularly vulnerable to climate change problems with, appropriately, water and agriculture as the first two, followed by biological diversity and coastal environment and communities.[28] These culminated in January 2011 with the long-awaited National Conference for the Conservation and Management of Water Resources in Yemen.[29] Its recommendations largely reasserted those of the National Water Sector Strategy and Investment Programme (NWSSIP) designed a few years earlier.

The outcome of all these conferences and documents was overtaken by events, first the popular uprisings throughout the country, followed by the ending of the Saleh era. The transition regime which was established in early 2012 drafted a Transitional Program for Stabilization and Development (TPSD) to cover the years 2012–2014, which focuses on security related issues, but did little to connect them with climate change or its effects, being pre-occupied by short-term political and security crises. As the transition unravelled and the country's state structures became increasingly ineffective from 2013 onwards, implementation of all social and economic policies, even those addressing the environmental urgency, moved to the remotest back burner of decision making while politics took an increasingly military turn and the power struggle took precedence over all other issues.

Hence, the TPSD is the most recent official policy document, and may be revived in the future. It has a significant section on the environment including, among others, studies 'to highlight and quantify the links among environmental degradation, methods of environmental exploitation, poverty, jobs and livelihoods loss, and population growth.'[30] On water, it merely reasserts the recommendations of the NWSSIP including 'mainstream and encourage community-based and grass root level integrated water resource management and partnership among government, community organisations, private sector companies etc.' and 'enhance the current processes of construction of water dams and water harvesting in terms of their ability to generate rural employment activities and village/community level water infrastructure.'[31] In other words it repeats a series of policies that have never been demonstrated to work or be appropriate and which have been seriously questioned by many experts working in the sector.

Addressing climate change

Given that 70 per cent of the population are rural and the majority are reliant on agriculture for much of their income, in addition to the fact that 80 per cent of the poor also live in rural areas, the policies implemented with respect to agriculture and any aspects of rural development are of prime importance to address climate change and improve food and water security. Regardless of stated strategies, until the end of the 20th century, international agencies focused their assistance on irrigated agriculture: not only did they finance large-scale spate irrigation projects, for example in the Tihama, Tuban and Abyan areas, but the vast majority of the research carried out by the Agricultural Research and Extension Authority (AREA) was on high value, export-oriented, irrigated crops.

It is only in the past decade, when the water emergency had become clear to even the least-educated Yemeni that any support has been given to the development of rainfed crops. Even this has been minimal and primarily through the joint World Bank and International Fund for Agricultural Development (IFAD) supported Rainfed Agriculture and Livestock Project (RALP) whose design was finalised in 2006 and which was completed in 2013. It is worth noting, however, that most funding for this project has been devoted to 'community development' investments rather than on identification, development and dissemination of drought-resistant and fast-maturing varieties of rainfed crops; of a total World Bank US$20 million loan for this project, only US$3.3 million was devoted to a 'farmer-based system of seed improvement and management.'[32] While many community development activities, such as *wadi* bank protection and terrace rehabilitation, have positive environmental and production outcomes, far too little funding and attention has been given to the enormous research and extension needs for drought resistant and fast maturing varieties of both staple and cash crops.

As mentioned earlier, international funding agencies have played a major role in determining the Yemeni government's policies, including those addressing climate change. The Washington Consensus policies implemented in so many countries have also featured in Yemen where various governments have resisted them to the extent possible by delaying imposed reforms and other similar measures. However, the state's dependence on foreign funding for investment costs and, all too often, running expenses, has meant that most development interventions have been far closer to the neo-liberal model than might have been the case had it had access to other sources of financing, including self-financing. Here, we will look in some detail at World Bank interventions as this institution is by far the most important one with respect to its influence on government. Others, whether UN agencies, other multilaterals like IFAD or most bi-lateral funders also follow similar strategies.

Out of the US$423 million approved by the World Bank for Yemen since 2011, about 10 per cent is allocated to explicitly climate-related investments, including the aforementioned as well as a contribution of US$20 million (from a total cost of US$140 million)[33] towards the financing of a wind park project in Mokha to produce 60 GWhe of clean electricity per year, as well as USD$2.61 million for

the development of biogas digesters. On behalf of the Climate Investment Fund, the World Bank is managing Yemen's participation in the Pilot Project for Climate Resilience, which includes three projects for a modest total amount of US$50 million,[34] none of which had started by early 2015. They are, first, Development of a Climate information system (US$20.9 million), and second, the Climate Resilience Integrated Coastal Zone management project (US$20 million). Finally, the Climate Resilience of Rural Communities has been under preparation in 2014; it closely resembles the 'community development' components of the RALP and is expected to receive US$9 million in grant and US$37 million as a concessional loan. All World Bank activities have been suspended since the war started; their resumption will only follow the establishment of an effective government controlling the country.

Implications of climate change

Yemen faces a very difficult future, regardless of its current political upheavals. A number of very basic environmental challenges need to be addressed with the utmost urgency if the country is to survive as a habitable environment. Its limited and diminishing water resources need to be managed with extreme care to enable the sustainability of both immediate human domestic and drinking needs and agriculture, though the latter will have to be on a reduced scale. Water scarcity also means that jobs requiring the minimum possible water will have to be created, which implies a fundamental improvement in the educational system to enable people to be qualified for such employment. Exhaustion of water in the highlands means that many rural people will have to migrate to the cities, particularly those on the coast which can benefit from desalinated sea water for basic human needs, something on which investment has, to date, been non-existent. In addition, these cities will also need to be re-designed to cope with rising sea water levels.

While Yemen is subject to regular, ongoing environmental hazards. Specific trigger events are the increasing numbers of droughts and floods, due to the variation in rainfall patterns and intensity, as well as the kind of emergencies that are likely to arise from rising sea levels. Those most exposed and vulnerable to environmental hazards are the rural poor and the coastal populations as was demonstrated so vividly by cyclone Chapala. While Yemenis have proved their resilience to a difficult environment over centuries, this is weakening particularly for the poor who don't have any protective buffer either in terms of savings or in the form of stronger bodies. It is always the poor who suffer first and most from disasters.

Poverty is the major symptom of Yemen's inability to provide for its citizens. With a poverty rate of 54 per cent or more in the 2010s, and with no signs of improvement, poverty is due not only to inappropriate state and international funding agencies' development policies, but also to the country's aridity and climate stresses. We have seen how shortages of water and soil deterioration reduce the rural population's ability to increase incomes from agriculture, while other factors such as low investment and low educational standards are additional factors

worsening poverty. In itself, poverty is a cause of insecurity as it is an incentive to unrest and criminality, including insurgency.

Food security cannot be achieved through the expansion of agriculture, though there is significant scope for improvement through the development of high yielding rainfed cash and staple crops, as could be achieved by the wide dissemination of efficient irrigation methods in some areas. This would make it possible to sustain at least some of the population in Yemen's beautiful scenic highlands. Firmly implemented water management regulations will be essential to retain as many smallholders as possible in rural areas. These include preventing the over-exploitation of deep aquifers with powerful pumps and deep boreholes to save the shallower aquifers for the smaller and poorer farmers. Regardless of effectively implemented measures, the Yemeni nation will remain extremely dependent on imported staples and other foods due to its limited national resources and rapidly increasing population. Hence, it is essential to develop an economy able to satisfy the population's basic needs, including the purchase of increasingly expensive food given the predicted international rise in food prices as a result of climate change and increased world populations.

Conclusion

Some experts are predicting that it will take 20 to 30 years to solve Yemen's political problems. Unfortunately, the country's people cannot wait that long for solutions to fundamental issues. In particular, if nothing is done, most of the country will have run out of water before then, thus dramatically reducing the area suitable for human habitation and thus the number of people who can survive in Yemen. Yemenis need immediate, firm political action if the country is to survive as an entity. This situation clearly reveals the triviality of current struggles and debates about southern separatism, Islamic fundamentalism, factional fights between the different sub-groups of the minuscule 'elite' or even those between Huthis and others. While the 18 Yemeni billionaires will simply migrate to their second or fifth homes in some other wealthy enclave, what will happen to the projected 45 to 50 million Yemenis in 2035?

In brief, what is needed urgently is an effective and strong democratic government committed to addressing the country's fundamental problem of resource management and the expected additional stresses of climate change. This must be done through policies that ensure that the population can achieve reasonable living standards locally in an environmentally sustainable manner, thus defusing the underlying causes of frustration and poverty that sponsor insurgency. The following policies must be given priority:

- management of the limited water supplies in a sustainable manner, giving priority to immediate human needs of domestic water, followed by livestock needs, industrial development and agriculture
- development of drought resistant and fast maturing varieties for staple and cash crops enabling rainfed and spate agriculture to be viable even on small holdings

- development of high yielding and high value livestock husbandry systems
- development of a quality education system to enable future generations to be involved in low natural resource–use, income-generating activities, thus reducing unemployment
- a complete range of mitigating measures to address climate change effects, in particular the new rainfall patterns, soil deterioration, and rising sea levels

This chapter has demonstrated that Yemen's fundamental long-term problems are closely related to the difficult environmental conditions of an arid climate, a situation which is expected to worsen in coming decades as a result of climate change, which has a number of features. Some of these, water shortages in particular, will affect the entire population and result in significant population movements. Combined with soil deterioration, water shortages will disproportionately affect the 11 million rural poor whose main source of livelihood will shrink and thus worsen their situation. Without urgent political solutions, reached through a constructive spirit and compromise, and the implementation of sustainable development policies, Yemen's future may well be one of mass emigration when the limited water supplies run out and political strife drives its population into becoming climate change and conflict refugees. The closure of all borders in the current war will not prevent desperate people from moving when they have no alternative.

Notes

1 An earlier version of this chapter was presented at the Fifth Gulf Research Meeting (GRM), 25–28 August 2014, which was organised by the Gulf Research Centre Cambridge at the University of Cambridge. Thanks to my colleagues on the 'Future of Yemen' Workshop for their constructive and helpful comments.

2 Detailed recent analyses of Yemen's contemporary problems can be found in H. Lackner, ed. (2014). *Why Yemen Matters*. Saqi, London and L. Bonnefoy, F. Mermier and Marine Poirier eds. (2012), *Yemen, Le tournant révolutionnaire*. CEFAS-Karthala, Paris.

3 Central Statistical Organization Annual Statistical Yearbook 1990.

4 Central Statistical Organization Statistical Yearbook 2012, p. 69.

5 OCHA, Humanitarian Dashboard as of December 31, 2015, published January 19, 2016.

6 Data from OCHA Yemen Humanitarian Bulletin 30, August–September 2014 and presentation by UN of the 2014 Humanitarian Response Plan in April 2014.

7 Saba News Agency, June 10, 2014.

8 World Bank, World Development Indicators 2015, based on UNESCO estimates in 2013.

9 See the Bureau of Investigative Journalism data on drone and other strikes in Yemen, on their 'Drone War' blog.

10 In 2009, it had a per capita consumption of KWH 190 compared to a MENA average of 1,418 and a world average of 2751KWh (World Bank (2014). *Mocha Wind Farms Project Appraisal Document*. Washington, DC. p. 2).

11 Ian Golding (2013). *Divided Nations*. Oxford University Press, Oxford. pp. 40–41, 44.

12 As discussed with respect to agriculture in M. Mundy and F. Pelat (2015). *The political economy of agriculture and agricultural policy in Yemen*. In Saud al Sarhan and Noel Brehony (eds.), *Yemen to 2020: Political, Economic and Social Challenges*. Gerlach Press, Berlin. Chapter 7.

13 The World Bank (2010). *Yemen: Assessing the Impacts of Climate Change and Variability on the Water and Agricultural Sectors and the Policy Implications*. 2010 report no 54196-YE, 95. Washington, DC.
14 Ibid. p. 20.
15 Matthew Weiss (2015). A perfect storm: The causes and consequences of severe water scarcity, institutional breakdown and conflict in Yemen. *Water International*, 40, 251–272.
16 For more details of my analysis of state policies, see chapter 8, Water scarcity: Why doesn't it get the attention it deserves? in Lackner, *op.cit*, pp 161–182.
17 World Bank (2010) ibid., p. 21.
18 See Chris Ward (2015). *The Water Crisis in Yemen.* I.B Tauris, London.
19 See Abbas El-Zein et al. (2014). Health and ecological sustainability in the Arab World: A matter of survival. *The Lancet*, 383, 458–476.
20 See Lackner, *op.cit*, p. 177.
21 Republic of Yemen, Ministry of Water and Environment (2005). *National Water Sector Strategy and Investment Program, 2005–2009.* Its update in 2008 is the latest official policy statement. Sana'a.
22 See James Firebrace (2015). Yemen urban water: Extreme challenges, practical solutions and lessons for the future: The case of Taiz. Saud al Sarhan and Noel Brehony (eds.), *Yemen to 2020: Political, Economic and Social Challenges*. Gerlach Press, Berlin. Chapter 6.
23 Marcus Moench (1999). *Yemen: Local Water Management in Rural Areas: A Case Study.* World Bank, Washington, DC. M. Moench and H. Lackner (1997). *Decentralized Management Study.* World Bank, Sana'a.
24 Government of Yemen website.
25 World Bank (2012). *Climate Investment Funds, Strategic Program for Climate Resilience for Yemen*. Meeting of the PPCR sub-committee, 17 April, p 42.
26 See Lackner, Chapter 8 and Lichtenthaler, Chapter 9 in Lackner, *op.cit*.
27 World Bank (1992). *Republic of Yemen: A Medium Term Economic Framework Report.* Yemen, 99172, p 58.
28 Republic of Yemen, Environmental Protection Authority. (2009). *National Adaptation Programme of Action*. World Bank, Sana'a. p. 5.
29 Christopher Ward, Naif Abu-Lohom and Suhair Atef, eds. (2011). *Management and Development of Water Resources in Yemen.* Sheba Centre for Strategic Studies, conference proceedings.
30 Ibid., p. 16 of annex 2.
31 Ibid., p. 17 of annex 2.
32 World Bank (2013). *RALP Implementation Status and Results and Project Appraisal Document*. World Bank, Washington, DC.
33 World Bank (2014). *op.cit.*
34 Yemen (2012). *Strategic Program for Climate Resilience, Prepared under the Pilot Program for Climate Resilience*. Sana'a.

References

Bonnefoy, L. and Catusse, M., eds. (2013). *Jeunesses Arabes du Maroc of Yemen: Loisirs, cultures et politiques*. La Découverte, Paris.

El Zein, A., et al. (2014). Health and ecological sustainability in the Arab World: A matter of survival. *Lancet*, 383, 458–476.

EM-DAT. (2014). The OFDA/CRED International Disaster Database – www.emdat.be – Université Catholique de Louvain – Brussels – Belgium.

Firebrace, J. (2015). Yemen urban water: Extreme challenges, practical solutions and lessons for the future: The case of Taiz. In al Sarhan and Brehony (eds.), *Rebuilding Yemen: Political, Economic and Social Challenges*. Gerlach Press, Berlin. 123–148.

Golding, I. (2013). *Divided Nations*. Oxford University Press, Oxford.

Harris, I., Jones, P., Osborn, T. and Lister, D. (2014). Updated high-resolution grids of monthly climatic observations–the CRU TS3. *10 Dataset International Journal of Climatology*, 34, 623–642.

Lackner, H., ed. (2014). *Why Yemen Matters*. Saqi, London.

Lichtenthaler, G. (2014). Customary conflict resolution in times of extreme water stress: A case study of a document from the northern highlands of Yemen. In Lackner (ed.), *Why Yemen Matters*. Saqi, London. 183–196.

Moench, M. (1999). *Yemen: Local Water Management in Rural Areas: A Case Study*. World Bank, Washington, DC.

Moench, M. and Lackner, H. (1997). *Decentralized Management Study*. World Bank, Sana'a.

Mundy, M. and Pelat, F. (2015). The political economy of agriculture and agricultural policy in Yemen. In al Sarhan and Brehony (eds.), *Rebuilding Yemen: Political, Economic and Social Challenges*. Gerlach Press, Berlin.

Republic of Yemen, Environmental Protection Authority. (2009). *National Adaptation Programme of Action*. Sana'a.

UNDP Yemen. (2011). *UN Development Assistance Framework for 2012–2015*. UNDP, Sana'a.

Weiss, M. (2015). A perfect storm: The causes and consequences of severe water scarcity, institutional breakdown and conflict in Yemen. *Water International*, 40, 251–272.

World Bank. (2010). *Yemen: Assessing the Impacts of Climate Change and Variability on the Water and Agricultural Sectors and the Policy Implications*. World Bank, Washington, DC.

World Bank. (2012). *Interim Strategy Note 2013–14*. World Bank, Washington, DC.

World Bank. (2013). *RALP Implementation Status and Results*. World Bank, Washington, DC.

World Bank. (2014). *Mocha Wind Farms Project Appraisal Document*. World Bank, Washington, DC.

Yemen. (2012). *Strategic Program for Climate Resilience, Prepared under the Pilot Program for Climate Resilience*. World Bank, Sana'a.

7 Climatic hazards in the Himalayan region

Prajjwal Panday

Introduction

Mountainous regions represent one of the most fragile and vulnerable regions on Earth owing to the intersection of steep gradients in topography, vegetation and climate. These regions thereby represent a key concern on the global climate change agenda (Beniston 2003; Kohler and Maselli 2009; Kohler et al. 2010). Mountainous regions occupy approximately 25 per cent of the global land surface and are home to ~26 per cent of the global population (Meybeck et al. 2001; Beniston 2003; Diaz et al. 2003). Hence, mountain regions are comprised of sensitive systems with great biogeophysical significance and that have gained recent attention owing to their vulnerability to climate change impacts (Beniston 2003; Messerli et al. 2004; Nolin 2012). The Hindu Kush-Himalayan (HKH) region and surrounding regions in South Asia epitomise such a mountain ecosystem at the forefront of global environmental change where cryospheric, hydrological and ecological processes are under threat owing to a warming climate. Although people have inhabited this region for centuries, coping with hazards and risks, there is growing scientific evidence that the HKH region has become increasingly hazardous and disaster-prone in recent decades, a process which could further intensify with climate change (Marty et al. 2009). In addressing climatic hazards in this region, we examine water-related hazards, extreme weather and climate, stress on water resources, agricultural production and food security, all of which amplify vulnerability of this region when they interact with societal and economical dimensions.

The HKH region sustains around 210 million people across eight countries (Afghanistan, Bangladesh, Bhutan, China, India, Myanmar, Nepal and Pakistan) with nearly 1.4 billion people depending on water across the major river basins such as the Indus, Ganges, Brahmaputra, Yangtze and Yellow River (Figure 7.1). The Indus, Ganges and Brahmaputra basins constitute one of the most agriculturally fertile regions globally, supporting the livelihoods of approximately 700 million people, 85 per cent of whom reside in lowlands (elevations less than 1000m) in India (79 per cent) and Bangladesh (18 per cent) (NRC 2012; Nepal and Shrestha 2015). Approximately 195 million people live within the Indus basin alone, which includes most of Pakistan. These basins provide ecosystems services such as drinking water, irrigation, hydropower and biodiversity, which are critical for the livelihoods of the growing population. It is further believed that the HKH region

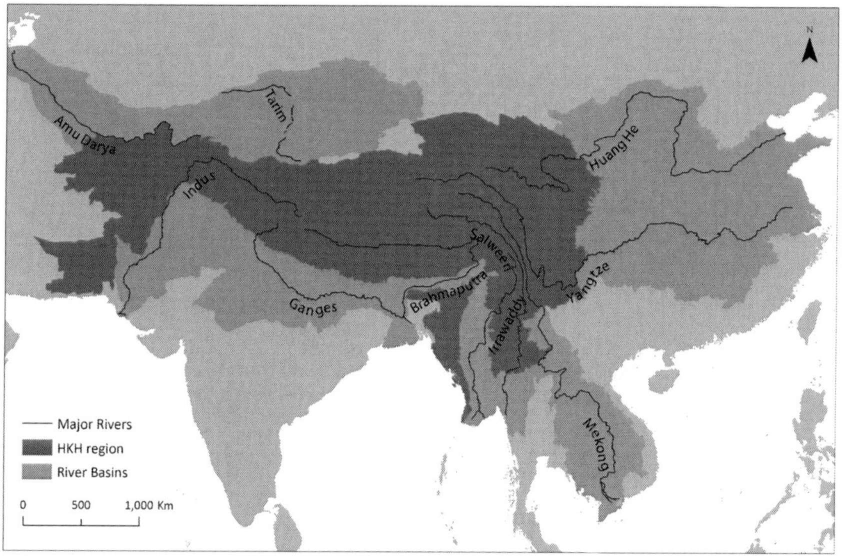

Figure 7.1 The Hindu Kush-Himalaya (HKH) region with the major river basins

Source: Pandey, P. (2016)

will increasingly face challenges in meeting water, food and energy demands in the future due to rapid urbanisation, population growth and socio-economic development (Mukherji et al. 2015). Business-as-usual scenarios for the region project that water demands and energy requirements need to double in order to meet additional food and energy requirements within the next decade (Molden 2007; Mukherji et al. 2015). On the other hand, the vulnerability of the people living in the HKH to natural and climatic hazards is exacerbated by poverty, marginality, limited accessibility and social fragility (Gerlitz et al. 2014). Furthermore, a changing climate with rising temperatures and precipitation changes may further push the hydro-climatological regime over critical thresholds thereby exacerbating climate-related hazards as well as water, food and energy scarcity (Lutz et al. 2014; Nepal and Shrestha 2015).

The HKH region represents a critical high-elevation environment in central Asia where cryospheric (frozen water) changes along with large-scale atmospheric, hydrological and ecological changes are already observable as warming temperature trends, glacier shrinkage and retreat, permafrost degradation, decreasing length of seasonal snow cover at higher elevations, earlier snowmelt runoff, degradation of grassland in the Tibetan Plateau, changes in Indian monsoonal patterns and local perceptions of a changing climate. Climate change impacts over the last several decades have been observed through significant temperature changes and also through variable rates of retreat of glaciers across the region (Bolch et al. 2012; Gardelle et al. 2012; Yao et al. 2012). Recent research analysing tropospheric temperatures reveals widespread annual warming rates over the entire HKH region of 0.21 ± 0.08°C/decade from 1979 to 2007 (Gautam et al.

2009, 2010). Even greater rates have been observed for the Nepalese Himalayan region specifically, with warming at approximately 0.6°C per decade from 1977 to 2000 (Shrestha et al. 1999). As the evidence suggests, potential impacts of a warming climate are greatest on regional hydro-climatology that may severely impact water resources in this region. Water-related hazards such as flash floods, outburst floods, landslides, debris flows and hazardous weather are projected to increase in the uplands. Recent flooding in the northwestern Himalayan region in the Uttarakhand State of India and western parts of Nepal exposed the vulnerability of these mountain regions to catastrophic flood disasters. Floods and cyclones are likely to increase in frequency, intensity and extent, and are expected to impact the lowland areas such as Bangladesh. At a basin scale, water availability and food security are also threatened by climate change owing to dependence on large-scale irrigation systems. This study provides analysis of climate-related hazards along with the susceptibility of harm and ability to cope of this region.

Climatic hazards and extreme events

Potential changes in climate have spurred a growing interest in climatic hazards and extremes that may have profound ecological as well as societal impacts globally (Easterling et al. 2000). One of the few available datasets on occurrences of hazards and disasters is the Emergency Events Database (Guha-Sapir et al. 2015; http://www.emdat.be), an International Disaster Database on disasters provided by the Center for Research on the Epidemiology of Disasters (Guha-Sapir et al. 2004). Data collection efforts have improved over the recent decades owing to better communication (Elalem and Pal 2014). According to EM-DAT hazard classification, a climatological hazard is usually caused by long-lived meso- to macro-scale atmospheric processes ranging from intra-seasonal to multi-decadal climate variability. Climate change impacts in mountainous regions can lead to hydrological hazards (flooding and landslides) and meteorological hazards (extreme weather events), all of which are most directly manifested through changes in extremes (Marengo et al. 2009; Huggel et al. 2010; Thibeault et al. 2010; Panday et al. 2014).

Hydrological hazards

Hydrological hazards in the Himalayan region are commonly related to hydro-meteorological conditions such as floods, landslides, avalanches, river-bank erosion and droughts (Pathak and Mool 2010). Extreme relief, enhanced orographic precipitation, thin soil over impervious bedrock and unstable mountain slopes contribute to the susceptibility of mountainous areas to hydrological hazards (Haritashya et al. 2006). Hazards such as flooding occur annually with the natural monsoonal cycle and are also positively correlated with the El Niño-Southern Oscillation phenomenon in the Indian subcontinent (Mirza et al. 2003). The frequency of reported annual occurrences of hydro-meteorological hazards across South Asian region indicates an increasing trend in the frequency of reported

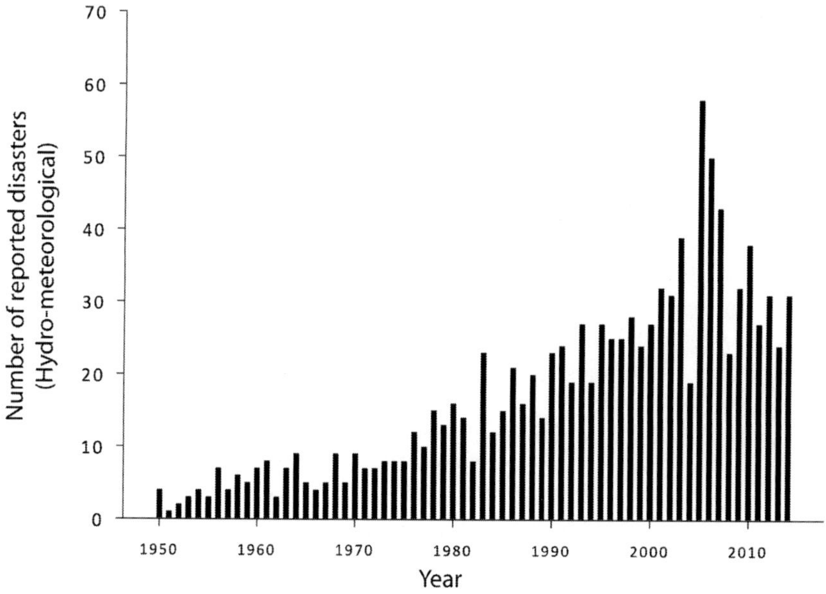

Figure 7.2 Number of hydro-meteorological disasters reported in the Emergency Events Database (EM-DAT) for South Asia (Afghanistan, Bangladesh, Bhutan, India, Nepal and Pakistan). Hydrological disasters include flood and landslides while meteorological disasters include extreme temperature and storms

Source: EM-DAT: The OFDA/CRED International Disaster Database – www.emdat.be – Université Catholique de Louvain – Brussels – Belgium

disasters (Figure 7.2). Although better data reporting may result in bias of number of reported disasters for recent decades, there is a possibility of the impact of climatic and socio-economic changes on the frequency of such extreme weather events and resultant disasters (NRC 2012; Elalem and Pal 2014). Projections based on Coupled Model Intercomparison Projects Phase 5 (CMIP5) simulations also indicate increased hydrological hazards such as flooding in South Asia (Mirza 2011; Hirabayashi et al. 2013). Climate change is expected to increase extreme precipitation events, shift the onset and departure dates of the summer monsoon, and affect the magnitude and frequency of cyclones, all of which may increase the frequency of hydrological hazards in this region (Krishnamurthy et al. 2009; Sen Roy 2009; Christensen et al. 2013).

Flooding

MONSOON FLOODING

The South Asian region is one of the most flood-vulnerable regions in the world. Flooding in the Himalayan region and the adjacent lowlands is commonly

associated with heavy or extreme rainfall events particularly during the Indian summer monsoon season or with artificial or natural dam bursts (Sen Roy 2009; Bookhagen 2010). The summer monsoon dominates the climate of eastern Himalayan region from the months of June and September and accounts for a large part of the annual precipitation budget over the Indian subcontinent. Subsequently, this results in flood events in the rivers and tributaries particularly in the Eastern Himalayan region. Long-term disaster data indicates that flooding events of various severities have mostly occurred over the last 30–40 years that is also consistent with the increase in intensity of the global monsoon (Webster et al. 2011). This data also indicates a high spatial heterogeneity in the occurrence of flooding disasters with northern regions of Afghanistan, Pakistan, India and Nepal in the western and eastern HKH showing high vulnerability to flooding (Elalem and Pal 2014).

Although total summer rainfall is far less in the western Himalaya and Indus River Basin, the severity of the monsoonal precipitation can sometimes be strong in these regions. Large flooding was observed in the Indus valley in Pakistan in the monsoon of 2010 (July–August) affecting 20 million people with 2,000 deaths and total damages exceeding US$40 billion (Houze Jr. et al. 2011; Webster et al. 2011). What started as heavy rainfall over a short period of time, leading to flash flooding, magnified into one of the worst disasters in the history of the country. Similarly extreme rainfall and flash flooding in the north Indian state of Uttarakhand in June 2013 led to massive landslides in the Mandakani River catchment, resulting in the loss of thousands of human lives and damage to property and infrastructure (Rao et al. 2014). The floods in the Kedarnath region in Uttarakhand were due to an unprecedented amount of rainfall over several days, averaging 360mm from June 15–18, 2013. Radar imagery of the region before and after this event indicates extreme erosion, channelization and deposition of sediments (Figure 7.3).

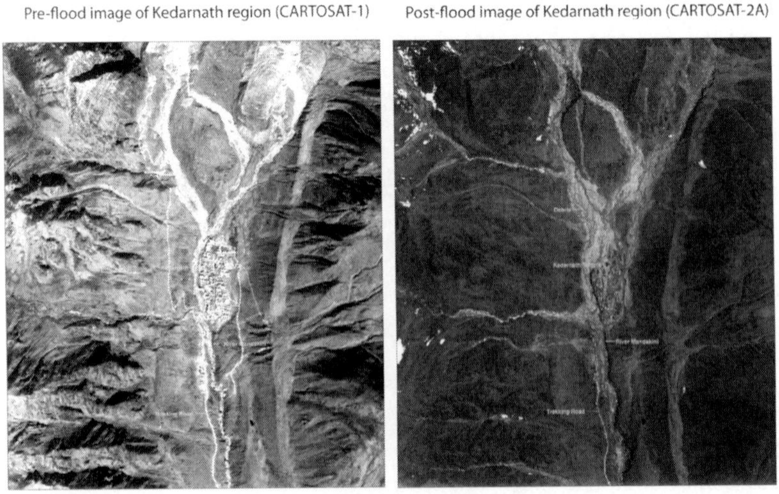

Figure 7.3 Radar imagery (CARTOSAT) of Kedarnath region taken before and after the flood event (NRSC 2015)

OUTBURST FLOODS

In upland areas on the high-mountain glacial environments, increased formation and expansion of ice- and moraine-dammed lakes has led to increased risks of glacial lake outburst floods (GLOFs, also jökulhlaup) in the HKH region (Richardson and Reynolds 2000; Mool et al. 2001; Bajracharya and Mool 2010). Greater warming at higher altitudes across the HKH region has led to the thinning and retreat of glaciers, accompanied by formation of new glacial lakes and expansion of pre-existing ones. An inventory based on remotely sensed satellite data mapping has identified around 2,400 glacial lakes in this region, of which several hundred are identified as potentially hazardous (ICIMOD 2011). Several outburst floods have already occurred in the past in this region, causing the loss of lives and property, and damage to infrastructure such as hydroelectric plants. Nepal alone has experienced around 14 GLOF events in the past. Recognition of such lakes as potentially hazardous has led to the installation of early warning systems and measures to lower lake levels in the Nepalese Himalaya (Benn et al. 2012). Glacial lakes have been shown to purge its melt-water within days, which is indicative of rapid melting. Figure 7.4 shows an image of glacial lake on the moraine of the Ngozumpa glacier in the Everest region of Nepal taken a day apart on May 28, 2013 and May 29, 2013, showing the level of drainage that occurred within a day. Recent studies have shown that sustained melting and ice loss in the decades to come owing to increasing temperature across high-altitude regions such as the Everest (Shea et al. 2014) will only accelerate the rapid formation of such potentially hazardous lakes (Thompson et al. 2012).

Landslides

Landslides are another common hazard in mountainous environments usually triggered by monsoonal rains, tropical cyclones and flooding in the Himalayan region (Hewitt 1997; Petley et al. 2007). Landslides are also causally linked to climate change through changes in precipitation (Huggel et al. 2012). Case studies from mountains of Europe, the Americas and the Caucasus indicate several mechanisms that can alter landslide magnitude and frequency under warming conditions which include the triggering of mass movement processes, slope instability due to permafrost degradation, tipping points in geomorphic systems and storage of sediment and ice involving important lag-time effect (Evans and Clague 1994; Gruber and Haeberli 2007; Huggel et al. 2012). Debris flows are common occurrences in mountain areas such as the Himalayas, but there is no clear evidence of increase in such events in peri-glacial environments. However, extreme rainfall events have increased in many regions of the world that commonly activate debris flows (Jones et al. 2007; Huggel et al. 2012). Landslide database for Nepal from 1978–2005 also indicates an upward trend in landslide fatalities despite high level of interannual variability in the occurrence of landslides (Petley et al. 2007). However, this increase in fatalities has been linked to rural road construction and deforestation as there are no substantial changes in rainfall patterns during the time period.

Figure 7.4 Rapid drainage of a glacial lake on the Ngozumpa glacier as shown by the drop in the water levels between May 28 and May 29, 2013 (Horodyskvj 2013)

Case study: Landslide in Sunkoshi, Nepal

In August 2014, heavy rainfall for several days caused a massive landslide in Jure village in Sindhupalchowk district of Nepal upstream of the Sunkoshi Hydropower project. The landslide led to the collapse of nearly 2km of hillside moving around 5.5 million cubic metres of debris that killed 156 people and displaced several hundreds. This landslide created a natural debris dam on the Sunkoshi River, forming a lake upstream which drained naturally without causing any fatalities. There have been three major flooding events in the Sunkoshi valley and its weak geological formation and steep topography combined with intense rainfall events, particularly during the monsoon, makes it vulnerable to hazards such as landslides. Figure 7.5 shows the extent of the landslide captured by Landsat 8, Operation Land Imager (OLI) images of the area on September 15, 2013 (pre-landslide) and September 2, 2014 (post-landslide).

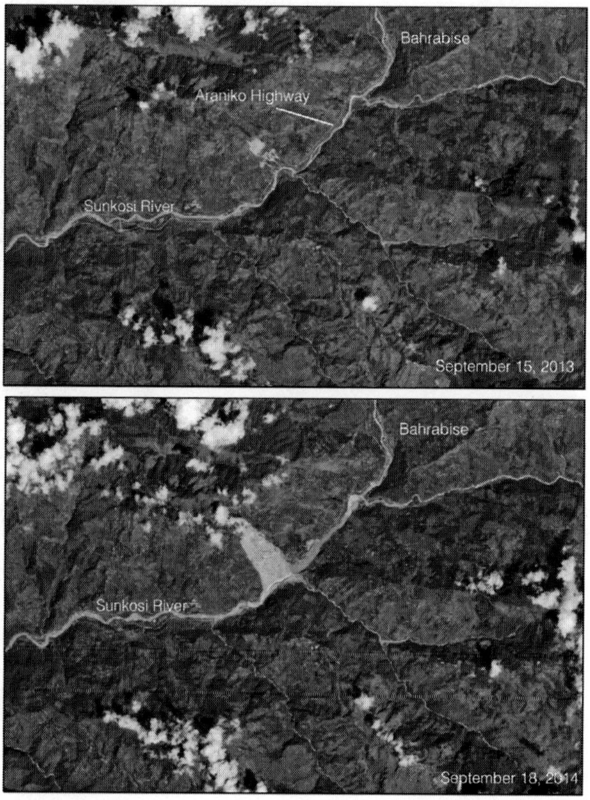

Figure 7.5 Landsat 8 images showing the Sunkoshi landslide area prior to the event in 2013 and post-landslide in September 2014 (NASA Earth Observatory 2014)

Extreme events

Observed historical trends of climate and future projections of climate change under different scenarios suggest that higher elevation regions will continue to experience the strongest warming across the planet (Beniston 2000; Beniston 2006; Déry 2011). This anomalous warming has prompted the attention to extreme weather and climate events due to the often large, catastrophic loss of human lives, the increasing economic costs and the disproportionately large part of climate-related damages associated with them (Easterling et al. 2000; Meehl et al. 2000). Such short-lived extreme weather events can strongly influence mass-transport processes and impact the rates of surface erosion processes (Bookhagen 2010). Klein Tank et al. (2006) and Panday et al. (2014) analysed station data for the late 20th century and show observed shifts towards climate extremes (fewer frost days and more warm nights) for the 1971–2010 across higher elevation areas in the Eastern Himalayan and South Asian region. At the same time, historical records of precipitation, streamflow and drought indices indicate increased drying over many land areas including South Asia since 1950 along with projections of severe drought over the coming decades (Dai 2011, 2013). Multi-model average projections from the Coupled Model Intercomparison Project (CMIP3 and CMIP5) indicate continued trends towards more extreme conditions consistent with a warmer, wetter climate in the Himalayan region, with more frequent temperature- and precipitation-related extremes, particularly for the Eastern Himalayan region (Panday et al. 2014).

The summer monsoon exerts a major influence on the regional hydro-climatology of South Asia. An understanding of its variability and underlying mechanisms of change is a fundamental challenge for climate science (Turner and Annamalai 2012). Long-term observations and climate models indicate a departure from normal monsoon years with a warming climate. Sen Roy (2009) analysed extreme hourly precipitation patterns in India from 1980 to 2002 and found rising trends in extreme heavy precipitation events, particularly in the high-elevation regions of the northwestern Himalaya as well as along the foothills of the Himalaya. The incidence of heavy monsoon rains has doubled from 1951–2000 as a result of significant rising trends in frequency and magnitude of extreme rain events (Goswami et al. 2006). Daily rainfall variability has increased with rain occurring less frequently but with more variability in intensity leading to increased frequency of both light and heavy rainfall events (Singh et al. 2014).

Glacier melt and streamflow

Geographic areas where the water cycle is dominated by snowmelt hydrology are expected to be more susceptible to climate change as it affects the seasonality of runoff (Barnett et al. 2005; Adam et al. 2009). Regional climate projections by the Intergovernmental Panel on Climate Change (IPCC 2007) indicate Central Asia may warm by a median temperature of 3.7°C by the end of the 21st century, with the greatest warming over higher altitudes (particularly the Tibetan Plateau

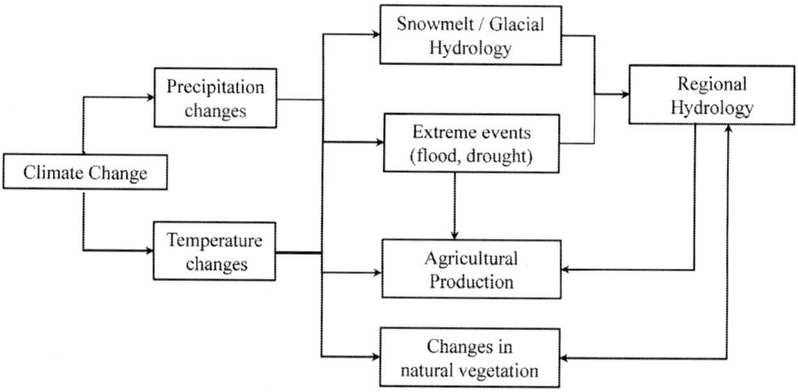

Figure 7.6 Flow diagram emphasising the connectivity and significance of regional hydro-climatology and climate change impacts to agricultural production for the HKH region (Panday 2013)

and the Himalayas). The HKH is one of the most glaciated regions outside of the polar regions occupying 60,054km^2 and an estimated ice reserve of 6,1247km^3 (Bajracharya et al. 2015). Glaciers across the Himalayas, except for the Karakoram Ranges, have retreated since the mid-19th century and mass loss has accelerated over the recent decades (Bolch et al. 2012; Bajracharya et al. 2015). There is an increasingly large body of evidence of a glacio-hydrological response along the east–west transect of the HKH corresponding to the these climatic changes, with glaciers in the eastern Himalaya exhibiting retreat and negative mass balance, and glaciers in the Karakoram and northwestern Himalaya exhibiting a positive mass balance over the last few decades (Panday et al. 2011; Bolch et al. 2012; Gardelle et al. 2012; Yao et al. 2012). Despite large variability in melt runoff regime, rising temperatures are projected to shift snow lines upward, shrink glacier coverage and reduce snow storage capacity thereby increasing melt runoff with greater hydrological extremes and shifts in seasonal peaks of total annual water availability (Lutz et al. 2014; Mukherji et al. 2015). These potential impacts of climate change, if realised fully, are greatest on the regional hydro-climatology which may severely impact water resources, irrigation, and hydropower generation (Figure 7.6).

Atmospheric hazards

Increasing concentrations of anthropogenic aerosols have negatively affected the air quality and climate over the Indo-Gangetic Plains (IGP) (Ramanathan and Ramana 2005; Gautam et al. 2010). Recent field experiments, in situ observations, and satellite monitoring have pointed out to the existence of atmospheric brown clouds (ABC) which are wide polluted tropospheric layers consisting of particles such as black carbon and sulphate aerosols from biomass combustion, power plants and vehicular pollutions (Bonasoni et al. 2010; Bonasoni et al. 2012;

Gautam 2014). These clouds can alter regional energy budget by reducing solar radiation at the surface by ~10 per cent, thereby reducing evapotranspiration and rainfall, while nearly doubling the atmospheric solar heating (Ramanathan et al. 2005). Previous studies have also found that adverse climate changes due to combined effects of ABCs and greenhouse gases have contributed to slowdown in rice harvest growth during the past two decades (Auffhammer et al. 2006).

During winter season, low temperatures, increased frequency of alternate low and high pressure systems, fine aerosols and valley-type topography of the IGP bounded by the Himalayas provide ideal conditions for fog and haze formation over this region (Gautam et al. 2007). The formation of fog starts in the latter half of December, extends over a stretch of ~1500 km in length and ~400 km in width and covers some parts of the region for more than a month (Ali et al. 2004). These annual occurrences of fog have been a recent phenomenon over India, and are responsible for trapping pollutants and poor visibility which disrupt transportation, leading to deaths from vehicular accidents (Hameed et al. 2000). Figure 7.7 shows the spatial distribution of the composite mean fog/low-cloud occurrences from 2000–2006 during winter season generated from Terra/MODIS data for aerosol and cloud properties (MOD04 and MOD06) (Gautam et al. 2007). The average number of foggy days over the six-year winter period is larger in the central IGP compared to the eastern and western regions. The clear demarcation along the foothills of the Himalayas is evident indicating the persistence of fog in the low topography/valley of the Ganges basin.

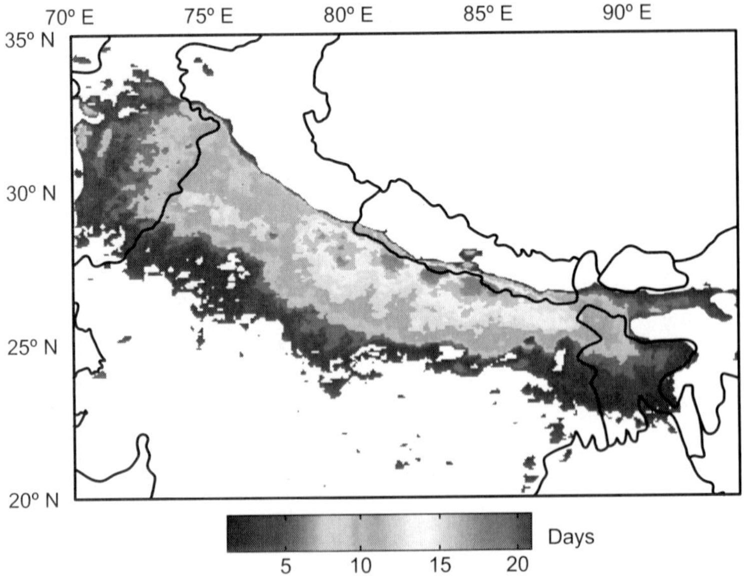

Figure 7.7 Composite mean fog/low-cloud occurrences for the six-year winter season (2000–2006) derived from MODIS cloud properties (Gautam et al. 2007)

Along with these events, the IGP is also exposed to high dust activity during pre-monsoon season from March to May when dust aerosols are transported from the Thar Desert in Northwestern India/Pakistan and the Arabian Peninsula (Gautam et al. 2009). Along with pollution implications, these dust storms can also have severe climatic effects. A mixing of this desert dust and soot aerosols over northern India and the Himalayan foothills may cause enhanced heating in the middle and upper troposphere over Tibetan Plateau, also known as the Elevated Heat Pump (EHP) effect, thereby strengthening the meridional tropospheric gradient and resulting in early advancement of the monsoon rainfall (Lau et al. 2006, 2008). Such potential interaction of aerosols and monsoon water cycle variability may have profound societal impacts of air pollutions and monsoon floods and droughts in South Asia (Lau et al. 2008).

Regional vulnerability

Impacts to agriculture and food security

One of the major potential impacts of climate change in this region is water availability and reduction of upstream discharges, which directly impact agricultural production and food security downstream (Immerzeel et al. 2010). Although very little land is irrigated in the upstream highlands of the HKH region, agriculture comprises the largest sector of water use in the lowlands in a majority of the countries within these larger basins. Irrigation is a major contributor to food security with 90 per cent of withdrawn freshwater used for irrigation in South Asia (Mukherji et al. 2015). By basins, the irrigated land areas are 149,900ha in the Indus, 156,300ha in the Ganges and 6000ha in the Brahmaputra (Nepal and Shrestha 2015). A large amount of irrigation is used for cotton production in the Indus basin, rice production in the Brahmaputra basin and primarily for wheat production in the Ganges basin (NRC 2012). As a result, the long-term sustainability of this region is dependent on water resource management, agricultural production and food security. Although South Asian countries increased food production during the early 1990s, production has slowed down since, leaving countries such as Afghanistan, Bhutan, Bangladesh, India and Nepal unable to cope with increasing population demands (Rasul 2010). Countries such as Afghanistan and Bangladesh are particularly vulnerable in terms of food security owing to increasing population, low productivity and lack of suitable land for arable agriculture. Datasets of hazards and disasters in this region have been dominated by occurrence of flood events in terms of the frequency of events and number of people affected; however, droughts and famine have accounted for more deaths in this region over the past century (NRC 2012). An analysis of climate risks for agriculture and crops in 12 food-insecure regions (based on statistical crop models and climate projections) identified South Asia as a region likely to suffer negative impacts on crops critical to large populations (Lobell et al. 2008). Given the poor food security conditions in this region and the surrounding countries, negative changes in water resources will certainly exacerbate these existing conditions.

Socio-ecological vulnerability of HKH

The vulnerability of this mountainous region to potential impacts from climate will increase over the next few decades. Vulnerability as defined by Turner et al. (2003) refers not only to the exposure to hazards alone, but also to the sensitivity and resilience of the individuals, groups or systems. The exposure of household and communities to livelihood stresses caused by both climatic and non-climatic factors, and their capacity to cope or recover, makes certain groups in a population disproportionality more vulnerable to disasters. Poverty analysis of member countries of the HKH, using the national living standard surveys, indicates that one third of the population is living in absolute poverty and, on average, there is 5 per cent higher poverty in the mountains than the national average of the respective HKH countries (Hunzai et al. 2011; Gerlitz et al. 2012; Shrestha et al. 2015). A regional study from the Nepalese Himalaya finds erratic rainfall, snowfall and prolonged drought as major climatic hazards that pose greatest threat especially to low-income farmers who lack capacity for short or long-term adaptation (Bhatta et al. 2015). Similarly, others have shown that, despite high exposure and sensitivity to climate change within different regions in Nepal, much of vulnerability is indicative of persistence of constraints on adaptation to local environments (Pandey and Bardsley 2015). A basin-scale vulnerability assessment by Varis et al. (2012) found the Ganges-Brahmaputra-Meghna (GBM) and Indus River basins to be the most vulnerable than any other river basins investigated in the Asia-Pacific region. The GBM basins also had the highest vulnerability with regards to vulnerability related to environmental hazards (Varis et al. 2012). The assessment of socio-ecological vulnerability is therefore critical for identifying climate adaptation requirements, disaster and hazards risk reduction measures, and post-disaster recovery efforts.

Transboundary water issues

Increasing energy demands of the region are certain to spur infrastructure development, particularly in construction of river dams to expand current hydropower capacity. The hydropower potential of the HKH region is estimated to be around 500 GW (Vaidya 2013) with 25GW of electricity generation potential alone in the Ganges Basin through upstream storage of water in proposed 23 dams upstream (Sadoff et al. 2013). Construction of all proposed dams in India would transform the Indian Himalaya to a region with the highest average dam densities driving the displacement of people and massive biodiversity loss driven by land use changes (Grumbine and Pandit 2013). China has also stepped up its efforts to harness hydropower with 750 planned projects planned in Tibet alone and potential diversion of water from the Tsangpo-Bramaputra (Bawa et al. 2010; Pomeranz 2013). The proposed construction of dams and/or inter-basin water diversion projects will impose major stress on riverine ecosystems and carry impacts downstream to lower riparian regions in India and Bangladesh (Kattelus et al. 2015). Lessons have been learnt from nearby Mekong River Basin. Here, rapid hydropower

development, with construction of a series of hydropower projects on the upper Mekong by China and 11 proposed dams in the lower Mekong, have already altered the ecological productivity of downstream ecosystems, such as the Tonle Sap, and may have further negative impacts to the downstream hydrological and floodplain regime (Grumbine and Xu 2011; Arias et al. 2014). The water resources are already overstretched in the South Asian region. Increasing interests in developing hydroelectric potential by both China and India will fuel the geopolitical tensions in the region associated with dams and transboundary water issues owing to unequal and unfair distribution of costs and benefits and governance issues (Kattelus et al. 2015).

Discussion and conclusion

The Himalayan region is under scrutiny in light of an increasingly large body of evidence of cryospheric, hydrologic and climatic changes. There is mounting warning signs that changes in regional hydro-climatology are inevitable and that the region is susceptible to increasing extreme climate and weather events and changing temperature and precipitation regimes in the coming decades. Although there is strong evidence that frequency of hazards across the Himalaya and surrounding lowlands has been increasing over the last decade, the linkage between climate change and hazards remains to be fully understood and investigated. Hazards and disasters in mountainous environments such as the Himalayas are also related to several other factors such as natural causes (e.g. seismic activity), environmental degradation and land use/land cover changes. Climate change is one of the many elements in this complex system that could amplify existing stress on resources, socio-economic, and political security.

A better understanding of the future trajectory of climate and the impacts of climate change on hazards, weather extremes and water resources is critical for adaptive and mitigation measures, and disaster preparedness and reduction. Current ability to predict future hazards and losses are particularly limited for this region mostly due to uncertainties associated with climate change projections. Despite the uncertainties, multi-model projections show agreement towards an overall increase in annual precipitation in the HKH river basins, reduction in snowmelt, increase in glacier melt and increase in climate extremes (Panday et al. 2014; Nepal and Shrestha 2015). Projections of demographic composition of specific regions are also limited for the region, but overall increased population and urbanisation translates to increased human exposure to climate-related hazards. We know that natural and hydro-climate hazards are not unique to this region and there is a high probability these may likely exacerbate given the inevitability of changes to the climate and hydrological system.

Recent hydrological disasters such as the Indus basin and Kedarnath flooding events have exposed the vulnerability of this region and also raised questions about its changing hydro-climatology, and the long-term implications for perennial and pervasive hazards in this region (NRC 2012). This gap in the knowledge of short and long-term implications of the impact of climate change on hazards in

the Himalayas is also due to lack of adequate, long-term monitoring and data collection efforts. Such data is critical for monitoring and management of resources, forecasting and early warning systems, as well as disaster mitigation and response. For better anticipation of future hydro-climatic hazards and disasters in this region, there are several challenges which include (a) high uncertainty in measurement, modelling and forecasting; (b) high spatio-temporal variability in hazard losses across multiple scales and (c) better understanding of socio-economic vulnerability and resilience (NRC 2012). The environmental security of the HKH therefore rests on regional cooperation amongst HKH member countries; information sharing; capacity building; improved monitoring, forecasting and warning systems and measures for disaster risk reduction.

References

Adam, J. C., Hamlet, A. F. and Lettenmaier, D. P. (2009). Implications of global climate change for snowmelt hydrology in the twenty-first century. *Hydrological Processes*, 23, 962–972.

Ali, K., Momin, G., Tiwari, S., Safai, P., Chate, D. and Rao, P. (2004). Fog and precipitation chemistry at Delhi, North India. *Atmospheric Environment*, 38(25), 4215–4222.

Arias, M. E., Cochrane, T.A., Kummu, M., Lauri, H., Holtgrieve, G. W., Koponen, J. and Piman, T. (2014). Impacts of hydropower and climate change on drivers of ecological productivity of Southeast Asia's most important wetland. *Ecological Modelling*, 272, 252–263.

Auffhammer, M., Ramanathan, V. and Vincent, J. R. (2006). Integrated model shows that atmospheric brown clouds and greenhouse gases have reduced rice harvests in India. *Proceedings of the National Academy of Sciences*, 103(52), 19668–19672.

Bajracharya, S. R., Maharjan, S. B., Shrestha, F., Guo, W., Liu, S., Immerzeel, W. and Shrestha, B. (2015). The glaciers of the Hindu Kush Himalayas: Current status and observed changes from the 1980s to 2010. *International Journal of Water Resources Development*, 31(2), 161–173.

Bajracharya, S. R. and Mool, P. (2010). Glaciers, glacial lakes and glacial lake outburst floods in the Mount Everest region, Nepal. *Annals of Glaciology*, 50(53), 81–86.

Barnett, T. P., Adam, J. C. and Lettenmaier, D. P. (2005). Potential impacts of a warming climate on water availability in snow-dominated regions. *Nature*, 438(17), 303–309.

Bawa, K. S., Koh, L. P., Lee, T. M., Liu, J., Ramakrishnan, P., Yu, D. W., . . . Raven, P. H. (2010). China, India, and the environment. *Science*, 327(5972), 1457–1459.

Beniston, M. (2003). Climatic change in mountain regions: A review of possible impacts. *Climatic Change*, 59, 5–31.

Beniston, M. (2006). Mountain weather and climate: A general overview and a focus on climatic change in the Alps. *Hydrobiologia*, 562(1), 3–16.

Beniston, M., ed. (2000). *Environmental Change in Mountains and Uplands*. Oxford University Press, New York.

Benn, D., Bolch, T., Hands, K., Gulley, J., Luckman, A., Nicholson, L., . . . Wiseman, S. (2012). Response of debris-covered glaciers in the Mount Everest region to recent warming, and implications for outburst flood hazards. *Earth-Science Reviews*, 114(1), 156–174.

Bhatta, L. D., van Oort, B.E.H., Stork, N. E. and Baral, H. (2015). Ecosystem services and livelihoods in a changing climate: Understanding local adaptations in the Upper Koshi,

Nepal. *International Journal of Biodiversity Science, Ecosystem Services & Management*, 11(2), 145–155.

Bolch, T., Kulkarni, A., Kääb, A., Huggel, C., Paul, F., Cogley, J. G., . . . Bajrachargya, S. (2012). The state and fate of Himalayan glaciers. *Science*, 336(6079), 310–314.

Bonasoni, P., Cristofanelli, P., Marinoni, A., Vuillermoz, E. and Adhikary, B. (2012). Atmospheric pollution in the Hindu Kush-Himalaya Region: Evidence and implications for the regional climate. *Mountain Research and Development*, 32(4), 468–479.

Bonasoni, P., Laj, P., Marinoni, A., Sprenger, M., Angelini, F., Arduini, J., . . . Biagio, C. D. (2010). Atmospheric brown clouds in the Himalayas: First two years of continuous observations at the Nepal Climate Observatory-Pyramid (5079 m). *Atmospheric Chemistry and Physics*, 10(15), 7515–7531.

Bookhagen, B. (2010). Appearance of extreme monsoonal rainfall events and their impact on erosion in the Himalaya. *Geomatics, Natural Hazards and Risk*, 1(1), 37–50.

Christensen, J. H., Kumar, K. K., Aldrian, E., An, S.-I., Cavalcanti, I.F.A., Castro, M. D., . . . Zhou, T. (2013). *Climate Phenomena and Their Relevance for Future Regional Climate Change*. Cambridge, United Kingdom.

Dai, A. (2011). Characteristics and trends in various forms of the Palmer Drought Severity Index during 1900–2008. *Journal of Geophysical Research: Atmospheres (1984–2012)*, 116(D12115).

Dai, A. (2013). Increasing drought under global warming in observations and models. *Nature Climate Change*, 3(1), 52–58.

Déry, S. J. (2011). Global warming and its effect on snow/ice/glaciers. In Singh, V. P., Singh, P. and Haritashya, U. K. (eds.), *Encyclopedia of Snow, Ice and Glaciers*. Springer Verlag, Dordrecht. 468–471.

Diaz, H. F., Grosjean, M. and Graumlich, L. (2003). Climate variability and change in high elevation regions: Past, present and future. *Climatic Change*, 59(1), 1–4.

Easterling, D. R., Evans, J., Groisman, P. Y., Karl, T., Kunkel, K. E. and Ambenje, P. (2000). Observed variability and trends in extreme climate events: A brief review. *Bulletin of the American Meteorological Society*, 81(3), 417–425.

Easterling, D. R., Meehl, G. A., Parmesan, C., Changnon, S. A., Karl, T. R. and Mearns, L. O. (2000). Climate extremes: Observations, modeling, and impacts. *Science*, 289(5487), 2068–2074.

Elalem, S. and Pal, I. (2014). Mapping the vulnerability hotspots over Hindu-Kush Himalaya region to flooding disasters. *Weather and Climate Extremes*, 8, 46–58.

Evans, S. G. and Clague, J. J. (1994). Recent climatic change and catastrophic geomorphic processes in mountain environments. *Geomorphology*, 10(1), 107–128.

Gardelle, J., Berthier, E. and Arnaud, Y. (2012). Slight mass gain of Karakoram glaciers in the early twenty-first century. *Nature Geoscience*, 5, 322–325.

Gautam, R. (2014). Challenges in early warning of the persistent and widespread winter fog over the Indo-Gangetic Plains: A satellite perspective. In *Reducing Disaster: Early Warning Systems for Climate Change*. New York, Springer. 51–61.

Gautam, R., Hsu, N. C., Kafatos, M. and Tsay, S. C. (2007). Influences of winter haze on fog/low cloud over the Indo-Gangetic plains. *Journal of Geophysical Research: Atmospheres (1984–2012)*, 112(D5).

Gautam, R., Hsu, N. C. and Lau, K. M. (2010). Pre-monsoon aerosol characterization and radiative effects over the Indo-Gangetic plains: Implications for regional climate warming. *Journal of Geophysical Research*, 115, D17208.

Gautam, R., Hsu, N. C., Lau, K. M., Tsay, S. C. and Kafatos, M. (2009). Enhanced premonsoon warming over the Himalayan-Gangetic region from 1979 to 2007. *Geophysical Research Letters*, 36, L07704.

Gerlitz, J.-Y., Banerjee, S., Hoermann, B., Hunzai, K., Macchi, M. and Tuladhar, S. (2014). *Poverty and vulnerability assessment – A survey instrument for the Hindu Kush Himalayas.* International Centre for Integrated Mountain Development, Nepal.

Gerlitz, J.-Y., Hunzai, K. and Hoermann, B. (2012). Mountain poverty in the Hindu-Kush Himalayas. *Canadian Journal of Development Studies/Revue canadienne d'études du développement*, 33(2), 250–265.

Goswami, B.N., Venugopal, V., Sengupta, D., Madhusoodanan, M. and Xavier, P.K. (2006). Increasing trend of extreme rain events over India in a warming environment. *Science*, 314(5804), 1442–1445.

Gruber, S. and Haeberli, W. (2007). Permafrost in steep bedrock slopes and its temperature-related destabilization following climate change. *Journal of Geophysical Research: Earth Surface (2003–2012)*, 112(F2).

Grumbine, R.E. and Pandit, M.K. (2013). Threats from India's Himalaya dams. *Science*, 339(6115), 36–37.

Grumbine, R.E. and Xu, J. (2011). Mekong hydropower development. *Science*, 332(6026), 178–179.

Guha-Sapir, D., Below, R. and Hoyois, P., 2015. EM-DAT: International disaster database. Catholic University of Louvain, Brussels, Belgium.

Guha-Sapir, D., Hargitt, D. and Hoyois, P. (2004). *Thirty Years of Natural Disasters 1974–2003: The Numbers.* Presses Univ. de Louvain, Louvain-la-Neuve, Belgium.

Hameed, S., Mirza, M.I., Ghauri, B., Siddiqui, Z.R., Javed, R., Rattigan, O.V., . . . Hussain, L. (2000). On the widespread winter fog in northeastern Pakistan and India. *Geophysical Research Letters*, 27(13), 1891–1894.

Haritashya, U.K., Singh, P., Kumar, N. and Singh, Y. (2006). Hydrological importance of an unusual hazard in a mountainous basin: Flood and landslide. *Hydrological Processes*, 20(14), 3147–3154.

Hewitt, K. (1997). *Regions of Risk: A Geographical Introduction to Disasters.* Longman, Harlow.

Hirabayashi, Y., Mahendran, R., Koirala, S., Konoshima, L., Yamazaki, D., Watanabe, S., . . . Kanae, S. (2013). Global flood risk under climate change. *Nature Climate Change*, 3(9), 816–821.

Horodyskvj, U. (2013). Ngozumpa Glacier Water Levels in May 28 and May 29, 2013. Retrieved 5 March 2015.

Houze Jr, R., Rasmussen, K., Medina, S., Brodzik, S. and Romatschke, U. (2011). Anomalous atmospheric events leading to the summer 2010 floods in Pakistan. *Bulletin of the American Meteorological Society*, 92(3), 291–298.

Huggel, C., Clague, J.J. and Korup, O. (2012). Is climate change responsible for changing landslide activity in high mountains? *Earth Surface Processes and Landforms*, 37(1), 77–91.

Huggel, C., Salzmann, N., Allen, S., Caplan-Auerbach, J., Fischer, L., Haeberli, W., . . . Wessels, R. (2010). Recent and future warm extreme events and high-mountain slope stability. *Philosophical Transactions of the Royal Society A: Mathematical, Physical and Engineering Sciences*, 368(1919), 2435–2459.

Hunzai, K., Gerlitz, J.Y. and Hoermann, B. (2011). Understanding Mountain Poverty in the Hindu Kush-Himalayas: Regional Report for Afghanistan, Bangladesh, Bhutan, China, India, Myanmar, Nepal, and Pakistan. International centre for integrated mountain development (ICIMOD), Kathmandu, Nepal.

ICIMOD. (2011). *Glacial Lakes and Glacial Lake Outburst Floods in Nepal.* ICIMOD, Kathmandu, Nepal.

Immerzeel, W. W., Beek, L.P.H. V. and Bierkens, M.F.P. (2010). Climate change will affect the Asian water towers. *Science*, 328, 1382–1385.

IPCC, 2007: Climate Change 2007: Synthesis Report. Contribution of Working Groups I, II and III to the Fourth Assessment Report of the Intergovernmental Panel on Climate Change [Core Writing Team, Pachauri, R. K and Reisinger, A. (eds.)]. IPCC, Geneva, Switzerland. 104

Jones, P., Trenberth, K., Ambenje, P., Bojariu, D., Easterling, D., Klein, T., . . . Zhai, P. (2007). Observations: Surface and atmospheric climate change. *IPCC, Climate Change*, 235–336.

Kattelus, M., Kummu, M., Keskinen, M., Salmivaara, A. and Varis, O. (2015). China's southbound transboundary river basins: A case of asymmetry. *Water International*, 40(1), 113–138.

Klein Tank, A.M.G., Peterson, T.C., Quadir, D.A., Dorji, S., Zou, X., Tang., . . . Sikder, A.B. (2006). Changes in daily temperature and precipitation extremes in central and south Asia. *Journal Geophysical Research*, 111, D16105.

Kohler, T., Giger, M., Hurni, H., Ott, C., Wiesmann, U., Wymann von Dach, S. and Maselli, D. (2010). Mountains and climate change: A global concern. *Mountain Research and Development*, 30(1), 53–55.

Kohler, T. and Maselli, D., eds. (2009). *Mountains and Climate Change: From Understanding to Action*. Geographica Bernensia and Swiss Agency for Development and Cooperation, Bern, Switzerland.

Krishnamurthy, C.K.B., Lall, U. and Kwon, H.-H. (2009). Changing frequency and intensity of rainfall extremes over India from 1951 to 2003. *Journal of Climate*, 22(18), 4737–4746.

Lau, K.M., Kim, M.K. and Kim, K.M. (2006). Asian summer monsoon anomalies induced by aerosol direct forcing: The role of the Tibetan Plateau. *Climate Dynamics*, 26(7), 855–864.

Lau, K.M., Ramanathan, V., Wu, G.X., Li, Z., Tsay, S., Hsu, C., . . . Chin, M. (2008). The joint aerosol-monsoon experiment: A new challenge for monsoon climate research. *Bulletin of the American Meteorological Society*, 89(3), 369–383.

Lobell, D.B., Burke, M.B., Tebaldi, C., Mastrandrea, M.D., Falcon, W.P. and Naylor, R.L. (2008). Prioritizing climate change adaptation needs for food security in 2030. *Science*, 319(5863), 607–610.

Lutz, A., Immerzeel, W., Shrestha, A. and Bierkens, M. (2014). Consistent increase in High Asia's runoff due to increasing glacier melt and precipitation. *Nature Climate Change*, 4(7), 587–592.

Marengo, J., Jones, R., Alves, L.M. and Valverde, M. (2009). Future change of temperature and precipitation extremes in South America as derived from the PRECIS regional climate modeling system. *International Journal of Climatology*, 29(15), 2241–2255.

Marty, C., Phillips, M., Margreth, S., and Kroup, O. (2009). Mountain hazards. In Kohler, T., and Maselli, D. (eds) *Mountains and Climate Change: From Understanding to Action*. University of Bern, Bern, Switzerland.

Meehl, G.A., Zwiers, F., Evans, J., Knutson, T., Mearns, L. and Whetton, P. (2000). Trends in extreme weather and climate events: Issues related to modeling extremes in projections of future climate change*. *Bulletin of the American Meteorological Society*, 81(3), 427–436.

Messerli, B., Viviroli, D. and Weingartner, R. (2004). Mountains of the world: Vulnerable water towers for the 21st century. *Ambio, Special Report 13*, 13, 29–34.

Meybeck, M., Green, P. and Vörösmarty, C. (2001). A new typology for mountains and other relief classes. *Mountain Research and Development*, 21(1), 34–45.

Mirza, M.M.Q. (2011). Climate change, flooding in South Asia and implications. *Regional Environmental Change*, 11(1), 95–107.

Mirza, M.M.Q., Warrick, R. and Ericksen, N. (2003). The implications of climate change on floods of the Ganges, Brahmaputra and Meghna rivers in Bangladesh. *Climatic Change*, 57(3), 287–318.

Molden, D. (2007). *Water for Food, Water for Life*. Earthscan, London and International Water Management Institute, Colombo.

Mool, P.K., Bajracharya, S.R. and Joshi, S.P. (2001). *Inventory of Glaciers, Glacial Lakes and Glacial Lake Outburst Floods: Monitoring and Early Warning Systems in the Hindu Kush-Himalayan Region Nepal*. Kathmandu, Nepal.

Mukherji, A., Molden, D., Nepal, S., Rasul, G. and Wagnon, P. (2015). Himalayan waters at the crossroads: Issues and challenges. *International Journal of Water Resources Development*, 31(2), 151–160.

NASA Earth Observatory. (2014). Before and after the Sunkosi Landslide. Retrieved July 2015.

Nepal, S. and Shrestha, A.B. (2015). Impact of climate change on the hydrological regime of the Indus, Ganges and Brahmaputra river basins: A review of the literature. *International Journal of Water Resources Development*, 31(2), 201–218.

Nolin, A.W. (2012). Perspectives on climate change, mountain hydrology, and water resources in the Oregon Cascades, USA. *Mountain Research and Development*, 32(1), 35–46.

NRC. (2012). *Himalayan Glaciers: Climate Change, Water Resources, and Water Security (0309260981)*. National Academies Press, Washington, DC.

NRSC. (2015). Kedarnath pre- and post-flood CAROSAT imagery. *National Remote Sensing Centre, Indian Space Research Organisation*. Retrieved February 9, 2015.

Panday, P.K. (2013). Hindu Kush-Himalayan Region at the Forefront of Global Change: An Assessment of Snowmelt, Hydrology, Vegetation, and Climate. PhD thesis. ProQuest Dissertations And Theses. AAT 3538198.

Panday, P.K., Frey, K.E. and Ghimire, B. (2011). Detection of the timing and duration of snowmelt in the Hindu Kush-Himalaya using QuikSCAT, 2000–2008. *Environmental Research Letters*, 6, 1–13.

Panday, P.K., Thibeault, J. and Frey, K.E. (2014). Changing temperature and precipitation extremes in the Hindu Kush-Himalayan region: An analysis of CMIP3 and CMIP5 simulations and projections. *International Journal of Climatology*, 35(10), 3058–3077.

Pandey, R. and Bardsley, D.K. (2015). Social-ecological vulnerability to climate change in the Nepali Himalaya. *Applied Geography*, 64, 74–86.

Pathak, D. and Mool, P. (2010). *Climate change impacts on hazards in the Eastern Himalayas*. International Centre for Integrated Mountain Development (ICIMOD), Kathmandu, Nepal.

Petley, D.N., Hearn, G.J., Hart, A., Rosser, N.J., Dunning, S.A., Oven, K. and Mitchell, W.A. (2007). Trends in landslide occurrence in Nepal. *Natural Hazards*, 43(1), 23–44.

Pomeranz, K. (2013). Asia's unstable water tower: The politics, economics, and ecology of Himalayan water projects. *Asia Policy*, 16(1), 4–10.

Ramanathan, V., Chung, C., Kim, D., Bettge, T., Buja, L., Kiehl, J., . . . Wild, M. (2005). Atmospheric brown clouds: Impacts on South Asian climate and hydrological cycle. *Proceedings of the National Academy of Sciences*, 102(15), 5326–5333.

Ramanathan, V. and Ramana, M. (2005). Persistent, widespread, and strongly absorbing haze over the Himalayan foothills and the Indo-Gangetic Plains. *Pure and Applied Geophysics*, 162(8–9), 1609–1626.

Rao, K. D., Rao, V. V., Dadhwal, V. and Diwakar, P. (2014). Kedarnath flash floods: A hydrological and hydraulic simulation study. *Current Science*, 106(4), 598.

Rasul, G. (2010). The role of the Himalayan mountain systems in food security and agricultural sustainability in South Asia. *International Journal of Rural Management*, 6(1), 95–116.

Richardson, S. D. and Reynolds, J. M. (2000). An overview of glacial hazards in the Himalayas. *Quaternary International*, 65–66, 31–47.

Sadoff, C., Harshadeep, N. R., Blackmore, D., Wu, X., O'Donnell, A., Jeuland, M. and Whittington, D. (2013). Ten fundamental questions for water resources development in the Ganges: Myths and realities. *Water Policy*, 15(S1), 147–164.

Sen Roy, S. (2009). A spatial analysis of extreme hourly precipitation patterns in India. *International Journal of Climatology*, 29(3), 345–355.

Shea, J., Immerzeel, W., Wagnon, P., Vincent, C. and Bajracharya, S. (2014). Modelling glacier change in the Everest region, Nepal Himalaya. *The Cryosphere Discussions*, 8(5), 5375–5432.

Shrestha, A. B., Agrawal, N. K., Alfthan, B., Bajracharya, S. R., Marechal, J. and van Oort, B. (2015). The Himalayan Climate and Water Atlas: Impact of Climate Change on Water Resources in Five of Asia's Major River Basins. http://lib.icimod.org/record/31180/files/HKHwateratlas_FINAL.pdf.

Shrestha, A. B., Wake, C., Mayewski, P. A. and Dibb, J. E. (1999). Maximum temperature trends in the Himalaya and its vicinity: An analysis based on temperature records from Nepal for the period 1971–1994. *Journal of Climate*, 12, 2775–2786.

Singh, D., Tsiang, M., Rajaratnam, B. and Diffenbaugh, N. S. (2014). Observed changes in extreme wet and dry spells during the South Asian summer monsoon season. *Nature Climate Change*, 4(6), 456–461.

Thibeault, J. M., Seth, A. and Garcia, M. (2010). Changing climate in the Bolivian Altiplano: CMIP3 projections for temperature and precipitation extremes. [D08103]. *Journal of Geophysical Research*, 115(D08103).

Thompson, S. S., Benn, D. I., Dennis, K. and Luckman, A. (2012). A rapidly growing moraine-dammed glacial lake on Ngozumpa Glacier, Nepal. *Geomorphology*, 145, 1–11.

Turner, A. G. and Annamalai, H. (2012). Climate change and the South Asian summer monsoon. *Nature Climate Change*, 2(8), 587–595.

Turner, B. L., Kasperson, R. E., Matson, P. A., McCarthy, J. J., Corell, R. W., Christensen, L., . . . Polsky, C. (2003). A framework for vulnerability analysis in sustainability science. *Proceedings of the National Academy of Sciences*, 100(14), 8074–8079.

Vaidya, R. (2013). Water and hydropower in the green economy and sustainable development of the Hindu Kush-Himalayan Region. *Hydro Nepal: Journal of Water, Energy and Environment*, 10, 11–19.

Varis, O., Kummu, M. and Salmivaara, A. (2012). Ten major rivers in monsoon Asia-Pacific: An assessment of vulnerability. *Applied Geography*, 32(2), 441–454.

Webster, P., Toma, V. E. and Kim, H. M. (2011). Were the 2010 Pakistan floods predictable? *Geophysical Research Letters*, 38(4).

Yao, T., Thompson, L., Yang, W., Yu, W., Gao, Y., Guo, X., . . . Pu, J. (2012). Different glacier status with atmospheric circulations in Tibetan Plateau and surroundings. *Nature Climate Change*, 2(9), 663–667.

8 China's natural and constructed hazard regimes

Troy Sternberg

Introduction

Throughout China's long history climate hazards have disrupted the society; in the last century >10 million people died in floods and in drought-related famines (Em-Dat 2015). The country's hazards are affected by geography, climate change and volatility, human activity and policy; these are factors that directly impact water and food supply, environmental productivity, governance and social stability. Three recent climate hazards exemplify the severity and range of climate disasters in China: the 2008 ice storms (77 million people affected, $21 billion damage), 2010 floods (134 million people, $18 billion damage) and the 2011 drought (525 million people, $94 billion damage) (Sternberg 2012; Em-Dat 2015). These represent disasters as physical, human and economic events. Equally, there is a dominant political role and latent moral hazard in how disaster is constructed in China. As the state enriches itself urban vulnerability is minimised whilst marginal rural populations, poor citizens, ethnic minorities and farmers and livestock raisers (900+ billion people) bear the risk of climate disasters (Chen et al. 2013).

China has great natural exposure to climate hazards and, since Mao's 'War on Nature', ever-increasing human-made hazard vulnerability (Shapiro 2001). Droughts, floods and storms are the totemic face of disaster; less obvious are the myriad ways human action increases exposure and alters natural environments. Direct and subtle shifts in policy, autocratic manias, economic trends, access to land and internal migration positions risk and places responsibility with marginalised individuals and communities (Chen et al. 2013). Further, the country's distinctive demographics, governance and landscapes draw together 1.3 billion people across diverse physical environments. Densely populated coastal zones, great rivers, extensive farmland, vast arid steppes and mountain plateaus expose the country to a range of disasters and differing levels of risk. At the same time, the autocratic government reconstructs the environment to meet perceived needs and policy agendas at multiple scales. This process favours development and economics over sustainability, natural resilience and conservation. A significant result is divided hazard realities where planning and disaster response have improved, yet vulnerabilities have often been rearranged rather than mitigated.

In all countries, natural hazards represent the interaction between a physical event and the social context in which it occurs. How land and resources are used, population dynamics and built environments frame hazard risk. Weather and extreme events are natural; their impact and damage depend on human systems. Recent international disasters reflect how societies engage, mitigate and exacerbate hazards across development levels. The great impact of 2005's Hurricane Katrina in the US is now attributed to the US Army Corps of Engineers poor river management (Rogers et al. 2015); the damage of Australia's extreme 2002–2007 drought reflected the exposure of farmers to climate conditions (Leblanc et al. 2009). Much of the impact of Pakistan's 2010 floods related to land control and management (see Rahim and Rueff, Chapter 3) whilst the 2013 Philippines typhoon reflected the placement of communities and floods in Taiwan highlight the influence of local practices in mitigating floods (see Dominelli, Chapter 11; Chen, Chapter 2). In this chapter, the three case studies discussed illustrate the way that natural disasters are linked to human activity in China. Today fake plastic trees are 'planted' in northern Gansu Province to stop wind and dust storms; the Three Gorges Dam impacts flood and drought in the Yangtze River basin whilst deforestation is a casual factor of drought and flooding in southwest Yunnan Province. Rather than a focus on hazards creating human risk, this chapter addresses how human systems create hazard risk in China.

When policy reconfigures nature, climate hazards take on political as well as physical and livelihood dimensions. Social processes, such as urbanisation, intensive agriculture, rapid development and resource degradation construct and exacerbate hazard scenarios as climate hazards reflect a duality of natural and human-influenced causes and outcomes. The conundrum is that, in a manipulated environment, human action can reduce hazard impact through preparation or strong response whilst simultaneously creating new risk parameters. For example, as communities mitigate drought through irrigation, new vulnerability is created through water shortages, wind and dust damage to fields, poorly built structures and the failure of unsuitable crops (assigned by government) in cold and dry landscapes (Sternberg 2014). China's natural and constructed hazard regimes are of great significance because of the country's economic strength, consumption of resources, as the source of water for 2 billion people and its centrality to the geo-politically critical Asia-Pacific region. Its global role makes China's hazard vulnerability and ensuing risk an issue of domestic and international significance as the well-being of a great number of people are at stake. To understand hazards, natural risk needs to be separated from the country's constructed disaster regime

Background

In the international arena, China is presented as a rising superpower, yet this masks the great disparity in wealth, landscapes and livelihoods within the country. For instance, 600 million people live on less than $4 day (World Bank 2015), exemplifying the country's high human risk profile. China has >20 per cent of the world's population but less than 7 per cent of global water in what is Asia's largest dryland

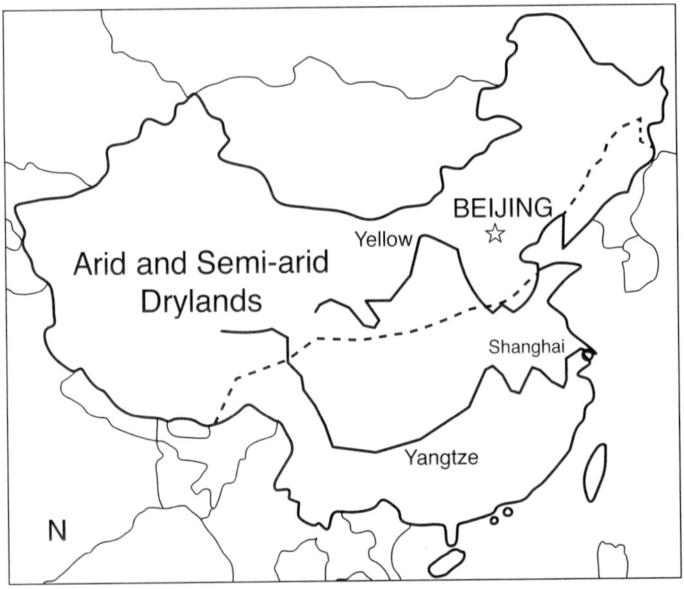

Figure 8.1 China, Yangtze and Yellow rivers, with drylands identified in the north and west (above dotted line)

nation (Sternberg et al. 2015) (Figure 8.1). This is further unbalanced as 60 per cent of the country's agriculture is produced in the northern plains that have 12 per cent of the water resources (Li 2012). As noted, policy stresses economic development over sustainability, a process reflecting the ruling party's focus on short-term economic goals as an effective way to maintain stability and political control. This has resulted in water scarcity, high pollution rates, tainted food supply and increased public health risks. At times the Yellow River no longer reaches the ocean, heavy metals contaminate >20 per cent of the country's soils and Beijing has recorded some of the world's highest air pollution levels (Wang and Hao 2012; Xinhua 2014). These extreme and immediate physical threats divert attention from unpredictable climate and hazard events.

China's great increase in infrastructure ($1 trillion invested in 2010) has led to the reconfiguration of landscapes, watersheds, environments and urban agglomerations that change and challenge physical dynamics and characteristics (Barreda and Wertime 2013). Efforts include transporting water 4,000 kilometres to Beijing, irrigating the desert for agriculture, rerouting water flow and damming rivers, farming cold and arid regions in the north and west to replace fertile land lost in the south and favouring policies for political, economic, ethnic or social reasons rather than environmental realities. The immediate impact creates new vulnerabilities as natural systems – flood plains, wetlands, windbreaks, drylands and soils – lose their functionality. At the same time, contrasting elements become

unnatural sights – sand dunes bordering crops, straightened rivers cutting new valleys, boats unable to operate on Asia's mightiest river – as mitigating ecosystems, including forests, fields, hillsides and open steppe become occupied spaces where previously cold weather, lack of water, exposure, poor soils and storms resulted in little human disruption.

Climate hazards are typically considered dramatic, random natural events caused by extreme weather; they become disasters when people are adversely affected. This is exemplified by severe drought, flood and storm episodes in China (Table 8.1). However, historically part of their great damage in China is a result of government policies and practices. Since the Peoples Republic of China was founded in 1949, the most famous disaster is often called 'Mao's Great Famine' of 1961–1964 (Bruins and Bu 2006; Dikötter 2010). This became a perfect storm of poor policy, drought, political insensitivity and failed harvests that resulted in millions (Em-Dat 2015) to tens of millions of deaths (Dikötter 2010). In the post-Mao era there has been greater awareness within the Communist Party that egregious human action can contribute to disaster-related mortality. A subtler approach would be to say the government has interest in reducing vulnerability or enhancing resilience to disasters as a measure to maintain social stability. Contributory risk factors include strong government control over land use and livelihoods at central, provincial and local levels, vast populations engaged in environmentally dependent livelihoods – 830 million farmers (FAO Stat 2011) – and limited ability for individuals to mitigate hazards through traditional methods such as livestock mobility, crop selection or access to water sources. Further restrictions on mitigating actions in today's context of development and infrastructure limit household and community response. As elsewhere, this leads to two different hazard environments resulting from natural and human-made disaster regimes in China.

Politics, population and economic priorities have instilled an ethos, particularly amongst policymakers, that nature is to be subjugated by man across China's large land mass (9.6 million km^2). Whilst the central and southern agricultural plains are well known, arid (<250mm mean annual precipitation) and semi-arid areas (<500mm m.a.p.) cover more than half of the country. This concentrates the

Table 8.1 Climate hazard events and impact in China, 1900–2010. EM-DAT 2015

Disaster	*Events*	*Deaths*	*People affected*	*Total damage (US$ Thousand)*
Drought	35	3,503,534	517,774,000	26,080,420
Flood	291	6,601,482	20,330,782,347	211,529,359
Storm	206	172,046	450,911,365	69,811,455
Temperature	13	379	81,220,002	21,430,200
Wildfire	6	265	56,616	110,000
	551	**10,277,441**	**3,080,687,714**	**302,851,434**

Note: hazard assessment methods vary – for example, recent accounts identity that half a billion were affected by the 2011 drought (Sternberg 2012)

population, development and risk in productive regions whilst limiting expansion and utilisation of vast areas of land due to physical constraints. Approximately 1.2 billion people live in central and southern China where many districts have >400–1000 people per square kilometre and high net in-migration rates (Atlas of China 2011). Other geographical areas include western and northern deserts with rural density <1–10 people per km²; the southwest (Tibet) is a high-altitude arid plateau and the northeast, a dry steppe grasslands. These regions have cold, continental climates where land has been increasingly converted to cultivation. In contrast, eastern and southern China are warm, densely populated temperate to subtropical zones that have resulted in one of the most human-altered landscapes in the world (Atlas of China 2011). Climatologically, droughts, wind and dust storms feature in the west, north and north central regions; floods and storms in the central and south and temperature extremes in the west and north: these dynamics contribute to geographical and human exposure to natural events and hazards.

Constructed disaster

Differentiating what is 'natural' vs what is 'human' is key to understanding extreme events in China. The concept of a 'constructed' hazard regime identifies anthropogenic (non-natural) forces and conditions that lead or contribute to disaster. Internally, this exemplifies typical 'Chinese characteristics,' or exceptionalism in the deliberate way government seeks to control, manipulate and disengage nature rather than accommodate or moderate activity to acknowledge the force of natural disasters in the built environment. China is a powerful example of this global trend because of its assertive belief in its ability to reshape nature to society's needs and desires. Exemplified by Mao's dogma that 'People Will Conquer Nature,' it represents the concept that humans have the ability (and right) to modify the physical world (Shapiro 2001). As case studies will show, this means placing farmers in the desert, damming and changing flow patterns of Asia's largest river, changing forests into monocultures and neutering natural mitigating systems. Follow-on remedies require cost-intensive engineering: reservoirs, canals and deep wells for farmers; desiltation, flood barriers and relocation of >1 million people on the Yangtze and anti-erosion measures, reforestation schemes and new drinking water sources in Yunnan. For climate hazards, the succinct point is that, as environments are reconfigured, exposure and adaptation to disasters will also change. The linear process can have positive or negative impacts on hazards; the risk environment is transformed and must be reinterpreted to recognise new challenges.

China's rapid development and rising living standards have been pursued with limited attention to, or integration of, environmental dynamics. The placement and expansion of towns, cities and industry and special economic zones, as well as the construction of dams on rivers in the Yangtze watershed, the conversion of marginal land to agriculture and the rapid urbanisation, drive related infrastructure. This is exemplified by the building of 1 million new highway kilometres from 1990–2010, the $77 billion South to North Water Transfer Scheme, adding 80–90 gigawatts of energy production annually (equivalent the UK's entire capacity) and

developing 50 new cities of 500,000 people represent the scale of infrastructure and change to the environment (Davidson 2014; Sternberg 2015). Concurrently, 2.6 million km^2 of land became desertified since 1990, 20 per cent of agricultural soil is contaminated with heavy metals and is not suitable for cultivation, 80 per cent of rivers and lakes are chemically polluted whilst 99 per cent of cities have worse air pollution rates than European minimum standards (Wang and Hao 2012; Wang et al., 2013a; Xinhua 2014).

Research identifies that development has converted 6–10 per cent of China's cropland to urbanisation, roads, industrial use and other infrastructure (Lin 2010). This includes land that is good for farming, beneficial as flood plains, provides soil stabilisation and water absorption, watershed services and similar natural hazard-reduction parameters. The lost agricultural production is replaced by adding and expanding marginal farmland in dry northern and western China (Lichtenberg and Ding 2008). Though this keeps the nominal amount of cropland constant, the environmental conditions and water scarcity make the land poorly suited for cultivation. Aridity, extreme cold and limited water expose northern farmland to drought, snow and ice disasters and wind and dust storms. This reduces productivity, affects livelihoods, spurs outmigration to cities and contributes to desertification in a self-reinforcing process. The outcome is increased vulnerability to natural disaster.

Within the constructed hazard regime government policy is the driving factor. Policies are motivated by growth targets and political discourse from the 'Great Leap Forward' to the 'Great Green Wall,' 'Sandification control for Beijing-Tianjin,' 'Convert cropland to forest or grassland' and similar programmes (Wang et al. 2013a); internal state restrictions (ecological resettlement, 'houkou' residence requirements); changing ethnic dynamics ('Go West' Han relocation, minority sedentarisation); special economic zones (multiple) and numerous economic and environmental control programmes. In this context, it is essential to separate nature, often identified as the causal event, from policy and socio-economic factors that frame and contribute to disaster. Thus the nation-building Three Gorges Dam, touted as a remedy to the flooding cycle on the Yangtze River, is also culpable for the extreme drought impact in 2011 when inadequate river flow meant a lack of water for irrigation and power generation, the inability of boats to travel on the river, loss of fishing and a partial drying of Pohang Lake, a key link in the South to North Water Transfer Scheme. Yet at the same time a strong government response prevented the damaged winter wheat harvest from spiralling into a disaster in the agricultural heartland, which is home to >500 million people (Sternberg 2012). This was done through monitoring and rapid response, $1.9 billion in water-related infrastructure, financial support to farmers and purchasing wheat on the international market to ensure adequate supply. The implied contradictions between natural vs constructed disasters make resolving the hazard contexts an ongoing struggle in China.

Fake plastic trees

Drought – we have no drought. We get our water from the ground.

(farmer, 38, Minqin, personal communication)

In Minqin County, Gansu Province, the Chinese are farming the desert – literally. The interesting dynamic is that since there is never enough rain for agriculture, drought is, in effect, a permanent condition. Cultivation is possible through a jumbled set of infrastructure that modifies the Tengger and Baidan Jaran region from an arid desert (115mm mean annual precipitation) to a marginally productive farming area home to 300,000 people. Key to understanding Minqin is Mao's desire to tame and control the border lands through Han in-migration (Zhang et al. 2007; Lacy 2008), the settlement of nomads (Sternberg 2014) and effective local government campaigns that have extracted several millions of dollars from the central government to 'keep the desert from swallowing Minqin.' These changes mean that farming has replaced pastoralism and livestock raising as the dominant livelihood on the dry steppe. The ever-increasing demand for water for domestic, agricultural, industrial and urban use in Minqin (Ma et al. 2005) suggests that the current system of swipe cards for water access, bureaucrat-driven crop selection for profit rather than water efficiency and the notion that farming will stop sand dunes spreading appear unsustainable.

Minqin County's extreme climate – temperatures ranging from -30⁰C to +40⁰C, limited precipitation, often in the form of snow, high evapotranspiration rates and environment – poor soils, low organic matter, high wind speed and dust entrainment – challenge rural livelihoods and reflect an area poorly suited to intensive cultivation. Yet government policy and subsequent support enables farming in the desert and mitigates the region's major disaster – drought. However, disaster risk is displaced from precipitation to wind and dust at local and national scales. Government loans for new-built structures to cope with cold temperatures and low precipitation prominently feature greenhouses. These have dirt foundations and are enclosed by heavy plastic sheeting which often rips in spring windstorms, exposing and killing plants and damaging structures. The hardship is that farmers are indebted to the government to pay for the plastic greenhouse; after windstorms, there are few resources to rebuild the sheds. Debt remains with the individual; the farmer may then be displaced to an urban centre to seek work. The second disaster is that, once the steppe rangeland is ploughed for farming, the topsoil is exposed to windstorms. The natural effect is that soil particles are picked up by the wind, contributing to massive dust storms that blanket Beijing in spring (thus the Stop the Beijing-Tianjin dust storms programme), affect Korea and Japan and reach the west coast of the United States (Akata et al. 2007).

The Minqin Anti-Desertification Centre, one of the agencies responsible for mitigating local disasters, exemplifies the nation's unique approaches to hazards. One method used is planting fake plastic trees as a key tool to fight desertification, dust and windstorms. The manufactured trees are placed in neat (unnatural) rows by hand in the hope of disrupting wind patterns that entrain the soil. Labelled as a cost-effective method, the audacity to pass off plastic shrubs as an ecological solution to climate events at a desert research institute is not grounded in ecological reality. A second option was for local farmers to 'volunteer' time planting some of the 56 billion seedlings that comprise China's Great Green Wall programme (Wang et al. 2013a). This optimistic scheme faces serious environmental and

human challenges. The former meant that there was insufficient precipitation for trees to grow without being watered. Yet watering the sandy desert soils requires human attention, infrastructure and costs at a scale that local governments are not interested in providing to meet national targets. Prior experience with government programmes meant the farmers understood the project dynamics – whilst volunteer labour may have been unavoidable (personal irrigation allotment depends on a government swipe card), labourers engaged in 'weapons of the weak' (Xie 2014). On assigned work days, farmers planted seedlings upside-down, roots sticking in the air: 'Why not? The officials do not know the difference' (farmer, male, 46; personal communication). Such efforts reflect the inefficiencies and contradictions of China's top-down government-driven disaster engagement.

The mighty Yangtze

At 6,300 kilometres long, the great Yangtze River basin is home to more than 400 million Chinese, produces 35 per cent of the country's agriculture, 40 per cent of the industrial output and 40 per cent of its effluent (Chen and Li 2005). As the cradle of Chinese civilisation for thousands of years, the river is part of the country's self-conception. This usually presents positive factors like beauty, water and life – yet equally, threat and disaster brought by nature and physical forces are reflected in the destructive power of the river. This was highlighted by a strong storm in 2015 that capsised a passenger boat, killing >400 people (BBC 2015). The symbolic representation of the public and state are matched by intensive use of the river for drinking water, irrigation, hydroelectric power, fishing, transport, industry and tourism and the construction of 50,000 dams in the watershed (Yang et al. 2011). There is a long history of the river's unkindness to its residents; in the last century, several major floods occurred, including in 1954, 1998 and 2010. The most dramatic was in 1931 when 3.7 million people died as a result of flooding and post-flood famine (History 2009).

Tracing floods identifies a river basin prone to variable rainfall, inundation, siltation and flooding, conditions that are exacerbated by human action over centuries. Predominantly located in a subtropical, monsoonal zone, the river starts on the Tibetan Plateau and then courses down through central China to the Pacific Ocean. Whilst average rainfall is 1,100mm, the precipitation in the middle and upper reaches exceeds 1,600 to 2,000mm annually (Chen and Li 2005). With up to 80 per cent of rainfall occurring in August–September, the river's high-to-low terrain distribution, water volume, exposure to soil erosion, sediment transport and deposition can disrupt flow and drainage, facilitate geological disturbance, particularly in reservoirs, and increase siltation, thus reducing holding capacity and increasing water flows.

Geography, topography and climate make flood a major risk on the Yangtze; human activity greatly exacerbates flood vulnerability. At the centre of controversy is the Three Gorges Dam, approved by the National People's Congress in 1992 and operational in 2003. As the world's largest hydroelectric scheme, the Three Gorges Dam was constructed to provide hydropower generation, flood control, economic

development and to serve as a nation-building project (Yang et al. 2011). It also displaced >1 million residents, lead to siltation and deforestation, required flood barriers and cost $37 billion to build (Lubin and Schafer 2010). In flood conditions, such as in May 2010, river siltation had reduced reservoir capacity and thus the ability to cope with great water volume. As reservoirs, which stretch ~500 kilometres and cover an area the size of Switzerland, rose precipitously more floodwaters had to be released, further inundating downstream regions (Wang et al. 2013b). Whilst the government claimed benefits of containment, other voices highlighted the heightened and unnatural downstream flows, land fragmentation, creation of new islands and industrial effluent and raw sewerage that make the water undrinkable (Stone 2008; Davies 2013).

Additional research identifies low discharge capacity, channel erosion and inadequate storage as contributory factors to flood damage in a region where river courses and reservoirs silt up quickly (Chen and Li 2005; Wang et al. 2013b). Operating on 10 to 20 years, the 'return flood' period has proved inadequate, particularly when other countries, such as Japan, plan for 1 to 2 hundred year events when designing flood control systems (Chen and Li 2005). Additionally, the damage is also water-borne diseases due to high sewerage levels, destabilised hill slopes in the reservoir that holds 39 billion m^3 of water, human displacement, potential seismicity and, if the dam were breached, 'one of the worst disasters in world history' (Stone 2008, p. 632).

The time between 2010 and 2011 was key for climate disasters on the Yangtze River. Severe flooding in August and September 2010 affected 230 million people and caused $50 billion in damage (Davies 2013). Much of the flood volume below the Three Gorges Dam was a result of excessive water in the reservoir that rose as much as 4m in a single night. The water level almost reached the maximum height of 175 metres in part because sediment in-fill reduced the capacity of the dam (Davies 2013). This threat led to a controlled release of floodwater over spill gates, increasing the water volume on the already saturated lower reaches of the river. As in past floods, the lower river's flatness (slope of 0.028/m/km) and siltation affected drainage and damage impact.

This was quickly followed by extreme drought in winter and spring 2010/2011 along the river's 1,600 kilometres below the Three Gorges Dam. The event was the worst drought since the People's Republic was formed, equivalent to a 1 in 100 year episode, and affected half a billion people (Sternberg 2012). Extremely low water levels meant there was insufficient water for irrigation (dessicating crops), it prevented boat transport, it dried fish ponds and river catch, it made hydropower generation unfeasible and it left millions of people without access to drinking water. Further, Pohang Lake, a key point in Beijing's South to North Water Transfer Scheme, fell to 30 per cent capacity, meaning the extensive dry lake bed would have had no water to reallocate to Beijing. The instigating causes were climate conditions, yet the impact was exacerbated by changing river flow, water allocation and human activity resulting from the Three Gorges Dam.

Yunnan's conundrum

Located in southwest China and abutting the Tibetan Plateau, Yunnan recently experienced its worst drought in decades as well as great flooding (Zhang et al. 2012). In 2009, drought covered >70 per cent of the region, resulting in a 60 per cent drop in precipitation that left 8 million people short of drinking water and led to crop failure, costing billions of dollars (Qiu 2010); that same year, a flood event affected 39 million people in the region (Em-Dat 2015). These events may appear to be acts of nature, yet they also reflect human action. They represent how 'climate change and poor environmental management can create a disaster' (Qiu 2010, p. 142). Changes in farming practices, environmental policy and engagement, deforestation and development all played roles in increasing vulnerability.

For example, 96 per cent of the tropical rainforest in Xishuangbanna Prefecture was removed between 1976 and 2003 (Li et al. 2008). The indigenous canopy was replaced by rubber trees, known as 'water pumps' for the amount of water they consume, and now cover 20 per cent of the prefecture (Qiu 2010). In other districts, eucalyptus, planted for fast growth, consume much water. The arboreal monocultures have developed alongside logging, mining and urbanisation that reflect the role of human activity in reconfiguring the hazard landscape. A natural result has been soil erosion and with it landslides and flash floods where once the rainforest provided natural protection. Deforested land loses its original hydrological role and productivity, thus when drought occurs forest fire risk increases whilst flood risk expands in wet, monsoonal periods.

Natural disasters offer a point of reflection, an opportunity to reassess management practices. In Yunnan, the challenge is in part an antiquated water delivery system serving expanding populations. A second issue is a dearth of alternate water sources in a region where many natural lakes, such as Dian Lake bordering the capital Kunming, are too polluted for human use. Further, whilst large infrastructure may be unfit for purpose or go unused, smaller reservoirs or distribution methods (canals, pipes) are inadequate to meet growing needs (Qiu 2010). Indeed, massive dams and reservoirs built to generate hydropower are driven by economic and political factors rather than for social or agricultural benefit or hazard reduction. The 2009 drought identified shortcomings in the water delivery system in a way that involves significant human and economic cost; nevertheless, in 2011, the government announced a $63 billion plan to build four more mega-dams to increase hydropower in the greater region (Qiu 2012). This is a reflection of the political power of the dam building industry, the perceived need for energy and the 'develop first' ethos over improving the effectiveness of existing infrastructure, reducing demand or increasing resilience.

Hazard preparation and response

The Yangtze, Yunnan and Gansu events present endemic disasters and reflect the government's variegated engagement with climate hazards. Efforts to mitigate disaster were showcased by rapid government response to the 2011 drought that

limited damage through intensive relief assistance, developing water resources in affected areas, financial support and, paradoxically, by allowing criticism of the Three Gorges Dam to relieve social pressure on the government (Sternberg 2012). Disaster management is coordinated by the China National Committee for Disaster Reduction, which includes no less than 32 agencies ranging from the Meteorological Administration, Ministries of Civil Affairs, Environmental Protection, Agriculture and Science and Technology to the surprising Radio, Film and Television, Publicity Department of the Central Committee, Public Security, Armed Police Force and People's Liberation Army Headquarters (Kang 2015). The convoluted composition and structure makes disaster management the responsibility of several contradictory and competing agencies and does not include the Emergency Management Office of the State Council, China National Committee for the International Decade for Natural Disaster Reduction, the National Disaster Reduction Committee or the Academy of Disaster Reduction and Emergency Management (Shi 2012). The inferred political control of several diverse agencies presents an unusual approach to disaster engagement. With more internal security directorates (police, etc.) than environmental agencies on the National Committee hazards instigate climate-driven human activity and mechanisms of control as much as mitigation of climate disasters. When decisions are taken for political expediency, particularly preservation of state power and social stability, hazards become 'staged' events, are converted to public relations opportunities and serve as justification for further state involvement to manipulate natural variability.

Against the backdrop of intensive management, several physical hazard parameters are challenging or beyond control. Of contemporary threats the greatest are climate change and variability, potential hazard implications for food and water systems, effective environmental coping mechanisms and the ability to manage and reduce modern stressors (pollution, contaminants, population, demand). As severe climate events affect huge populations, such as >500 million people in 2011's drought, manipulated landscapes can reconfigure and exacerbate risk; this is aggravated by unconstrained human demand for ecosystem services that have natural limits, particularly water and land (Sternberg 2012). China, as all countries, will continue to have climate hazard disasters; what is shifting is the distribution of causation and impact from natural forces to human-hazard influences to anthropogenic factors.

Counterfeit hazards

Cultural, social and structural realities further complicate hazard engagement. Issues such as unchecked government power, corruption, weak rule of law, regionalism, economic pressure and questionable data are major challenges. Disaster events can be opportunities for enrichment – in 2015, $14.4 million in disaster relief was diverted to bureaucrats by local governments in Shaanxi Province; after the 2008 quake, researchers at Tsinghua University identified that 80 per cent of donations went as 'extra revenue' to the government (BBC 2013; Economist 2015). Pervasive 'guanxi,' or personal networks and interwoven favours, can

circumvent legal structures, subvert policy and lead to corruption. These factors contribute to a contradictory approach to disaster events, the marginalisation of large segments of society (often rooted in ethnicity, poverty, livelihood or location) and foster a short-range outlook that neglects long-term risk perspectives and procedures. Issues of inequality, impunity and injustice lead to public cynicism and questions government intent. The outcome is selective engagement with hazards; the 2011 drought affecting national food supply and the Yangtze River Basin received much attention whilst government policy enabling farming the Minqin desert and deforestation in Yunnan went unaddressed.

Globally, 95 per cent of hazard-reduction funding is spent after the event occurs (Hochrainer and Mechler 2011). China follows this pattern at significant societal exposure as >20 per cent of the world's population faces significant disaster risk. This means that any event that disrupts water supply, food security, household livelihoods or economic activity will be writ large across Asia and the globe. It would be catastrophic, as Stone (2008) wrote, if a disaster spirals out of control in China – this would have unimaginable global consequences. Whilst the East Asian community focuses on China's economic power and military might, a potentially greater threat – a climate crisis – has received little attention. Whereas geopolitics makes financial, military and social issues problematic, if not impossible, to discuss between nations, physical threats, transboundary shocks and climate change at regional and global levels are serious, shared concerns. Several fora now exist for potential regional discussion and cooperation: the United Nations Framework Convention for Climate Change, Asia-Pacific Economic Cooperation climate symposia, Hyogo Framework Agreement on Disaster Reduction, U.S.-China Climate Change Working Group and the Association of Southeast Asian Nations (Rhee et al. 2015; Sternberg and Batbuyan 2013; Droge and Wacker 2014). These efforts can provide avenues for discussion of disaster scenarios and avenues for cooperation.

Chinese hazard dynamics are very much about climate events, human exposure, direct action and chosen response. As practiced, the process borders on moral hazard when the government chooses not to reduce risk but to follow policies and agendas that may increase vulnerability. This emphasises how risk that is disproportionally borne by marginal groups naturally exposed to climate – farmers and herders, rural and riverine residents, those with less access to state support – poor and minority groups and people without the connections to decision making and relief assistance. Multi-tiered hazard landscapes protect Beijing from climate impacts, cities and coastal communities are favoured over inland and countryside regions, those with access to wealth or *guanxi* are better insulated than the poor, and peripheral minorities (Tibetan, Uighur, Mongolian) are second to the dominant (Han) cultural group. Thus to address vulnerability requires a reading of social stratification and awareness of how policy, process and implementation diffuses across the vast country. Climate disasters are the ultimate external event filtered through the ingrained and doctrinaire Chinese system to great consequence.

Despite China's opaque governance structure, both the domestic and global community benefit from best practice in hazard preparation, mitigation and

response take place in the future. The threat of climate disasters is clear to the ruling Communist Party, yet the government struggles to reduce hazard exposure and, in several cases, displaces or increases risk through economic or policy-driven decisions that do not incorporate environmental forces or human vulnerability. Climate change already exacerbates hazards (IPCC 2014) whilst additional human dimensions, particularly those that are avoidable, create unnecessary exposure. The fear is that rather than being reduced, disaster risk will increase through human action. The hope is that China's 32 disaster management ministries and agencies will decrease vulnerability, improve response and mitigate risk for the country's 1.3 billion residents. The future will provide the answer.

References

Akata, N., Hasegawa, H., Kawabata, H., Chikuchi, Y., Sato, T., Ohtsuka, Y., . . . Hisamatsu, S. (2007). Deposition of 137Cs in Rokkasho, Japan and its relation to Asian dust. *Journal of Environmental Radioactivity*, 95, 1–9.

Atlas of China. (2011). National Geographic, Washington, DC.

Barreda, D. and Wertime, D. (2013). China's Great Infrastructure Binge, in Charts. www.theatlantic.com/china/archive/2013/08/chinas-great-infrastructure-binge-in-charts/278597/.

BBC. (2013). China's Red Cross fights to win back trust. http://www.bbc.co.uk/news/world-asia-china-22244339.

BBC. (2015). Yangtze Ship Disaster: Chinese Salvagers Right Eastern Star. www.bbc.co.uk/news/world-asia-china-33011557 (Accessed: 7 June 2015).

Bruins, H. and Bu, F. (2006). Food security in China and contingency planning: The significance of grain reserves. *Journal of Contingencies and Crisis Management*, 14(3), 114–124.

Chen, W., Cutter, S., Emrich, C. and Shi, P. (2013). Measuring social vulnerability to natural hazards in the Yangtze River Delta region, China. *International Journal of Disaster Risk Science*, 4(4), 169–181.

Chen, Y. and Li, J. (2005). Case Studies of Natural Disasters. *Natural and human induced hazards*. UNESCO, Paris, France.

Davidson, M. (2014). China's Electricity Sector at a Glance. www.theenergycollective.com/michael-davidson/335271/china-s-electricity-sector-glance-2013 (Accessed: 11 June 2015).

Davies, R. (2013). Flooding in China 2010. http://floodlist.com/asia/flooding-china-2010.

Dikötter, F. (2010). *Mao's Great Famine: The History of China's Most Devastating Catastrophe, 1958–1962*. Walker & Company, New York.

Dröge, S. and Wacker, G. (2014). China's approach to international climate policy: Change begins at home. http://www.ssoar.info/ssoar/handle/document/40187

Economist. (2015). Who Wants to be a Mandarin? www.economist.com/news/china/21653669-public-service-less-fun-if-you-cant-take-bribes-who-wants-be-mandarin (Accessed: 9 June 2015).

Em-Dat. (2015). www.emdat.be/ (Accessed: 1 April 2015).

FAO Stat. (2011). Market Intelligence on China's Agriculture & Food Industry. www.chinaag.org/production/china-agriculture/population-employment/ (Accessed: 22 August 2015).

History. (2009). Yangtze River Peaks in China. www.history.com/this-day-in-history/yangtze-river-peaks-in-china.

Hochrainer, S. and Mechler, R. (2011). Natural disaster risk in Asian megacities: A case for risk pooling? *Cities*, 28(1), 53–61.

Intergovernmental Panel on Climate Change (IPCC). (2014). Climate Change 2014: Impacts, Adaptation, and Vulnerability: Asia. Contribution of Working Group II to the Fifth Assessment Report of the Intergovernmental Panel on Climate Change. 1327–1370. Cambridge University Press, Cambridge.

Kang, Y. (2015). *Evolvement of Disaster Management Practices in China: A Trend Toward Greater Openness in Disaster Management in China in a Changing Era*. Springer, London.

Lacy, D. (2008). *The Chinese State at the Borders*. UBC Press, Vancouver.

Leblanc, M., Tregoning, P., Ramillien, G., Tweed, S. and Fakes, A. (2009). Basin-scale, integrated observations of the early 21st century multiyear drought in southeast Australia. *Water Resources Research*, 45(4), 1–10.

Li, H., Ma, Y., Aide, T.M. and Liu, W. (2008). Past, present and future land-use in Xishuangbanna, China and the implications for carbon dynamics. *Forest Ecology and Management*, 255(1), 16–24.

Li, S. (2012). China's huge investment on water facilities: An effective adaptation to climate change, natural disasters, and food security. *Natural Hazards*, 61, 1473–1475.

Lichtenberg, E. and Ding, C. (2008). Assessing farmland protection policy in China. *Land Use Policy*, 25(1), 59–68.

Lin, G. (2010). Understanding land development problems in globalizing China. *Eurasian Geography and Economics*, 51, 80–103.

Lubin, G. and Schafer, I. (2010). 17 Earthshaking Facts about the Three Gorges Dam and China's Next Even Bigger Water Project. www.businessinsider.com/three-gorges-dam-south-to-north-water-diverson-project-china-20107?op=1# ixzz3dishZyNh (Accessed: 22 August 2015).

Ma, J., Wang, X. and Edmunds, W. (2005). The characteristics of ground-water resources and their changes under the impacts of human activity in the arid Northwest China – A case study of the Shiyang River Basin. *Journal of Arid Environments*, 61(2), 277–295.

Qiu, J. (2010). China drought highlights future climate threats. *Nature News*, 465(7295), 142–143.

Qiu, J. (2012). Trouble on the Yangtze. *Science*, 336(6079), 288–291.

Rhee, J., Cai, W., Plummer, N., Sivakumar, M., Horstmann, N., Wang, B. and Kirono, D. (2015). Regional cooperation on drought prediction science for sisaster preparedness and management. *Bulletin of the American Meteorological Society*, 96(4), ES67–ES69.

Rogers, J. D., Kemp, G., Bosworth, H. and Seed, R. (2015). Interaction between the US Army Corps of Engineers and the Orleans Levee Board preceding the drainage canal wall failures and catastrophic flooding of New Orleans in 2005. *Water Policy*, 17(4), 707–723.

Shapiro, J. (2001). *Mao's War against Nature: Politics and the Environment in Revolutionary China*. Cambridge University Press, Cambridge.

Shi, P. (2012). On the role of government in integrated disaster risk governance – Based on practices in China. *International Journal of Disaster Risk Science*, 3, 139–146.

Sternberg, T. (2012). Chinese drought, bread and the Arab Spring. *Applied Geography*, 34, 519–524.

Sternberg, T. (2014). Transboundary hazard risk: The Gobi Desert paradigm. *Natural Hazards*, 72(2), 533–548.

Sternberg, T. (2015). Water megaprojects in deserts and drylands. *International Journal of Water Resources Development*, 32(2), 301–320.

Sternberg, T. and Batbuyan, B. (2013). Integrating the Hyogo Framework into Mongolia's disaster risk reduction (DRR) policy and management. *International Journal of Disaster Risk Reduction*, 5, 1–9.

Sternberg, T., Rueff, H. and Middleton, N. (2015). Contraction of the Gobi desert, 2000–2012. *Remote Sensing*, 7(2), 1346–1358.

Stone, R. (2008). Three Gorges Dam: Into the unknown. *Science*, 321, 628–632.

Wang, F., Pan, X., Wang, D., Shen, C. and Lu, Q. (2013a). Combating desertification in China: past, present and future. *Land Use Policy*, 31, 311–313.

Wang, J., Sheng, Y., Gleason, C. and Wada, Y. (2013b). Downstream Yangtze River levels impacted by Three Gorges Dam. *Environmental Research Letters*, 8.

Wang, S. and Hao, J. (2012). Air quality management in China: Issues, challenges, and options. *Journal of Environmental Sciences*, 24(1), 2–13.

World Bank. (2015). Poverty Headcount Ratio – Poverty Headcount Ratio at $4 a Day (PPP). http://databank.worldbank.org/data/home.aspx (Accessed: 16 December 2015).

Xie, Y. (2014). Institution, action and concept: Three dimensions of land expropriation dispute research. *Agricultural Science & Technology*, 15(6), 1054.

Xinhua. (2014). China Alerted by Serious Soil Pollution, Vows Better Protection. www.news.xinhuanet.com/english/indepth/2014-04/17/c_133270984.htm (Accessed: 26 February 2015).

Yang, S., Milliman, J., Li, P. and Xu, K. (2011). 50,000 dams later: Erosion of the Yangtze River and its delta. *Global and Planetary Change*, 75(1), 14–20.

Zhang, L., Xiao, J., Li, J., Wang, K., Lei, L. and Guo, H. (2012). The 2010 spring drought reduced primary productivity in southwestern China. *Environmental Research Letters*, 7(4), 045706.

Zhang, M., Borjigin, E. and Zhang, H. (2007). Mongolian nomadic culture and ecological culture: On the ecological reconstruction in the agro-pastoral mosaic zone in Northern China. *Ecological Economics*, 62, 19–26.

9 Temporal and spatial distributions of dust storms in Middle Asia

Natural and anthropogenic factors

Leah Orlovsky, Rodica Indoitu, Giorgi Kozhoridze, Madina Batyrbaeva, Irina Vitkovskaya, Batyr Mamedov and Nikolai Orlovsky

Post-Soviet Middle Asia is highly susceptible to dust storms due to its peculiar climatic conditions, including long rainless periods, large areas of sandy and clayey deserts, vegetation scarcity and frequent windstorms. Heavy dust storms sweep through Middle Asian countries every year and often adversely affect the natural environment and national economies. Dust storms aggravate soil erosion, blow seeds and emerging crops out of the ground, inflict damage on arable and grazing lands, hinder transportation and impact human health.

There are some discussions regarding the terms 'Middle' and 'Central' Asia. In the Russian (and later Soviet) literature, there has been a division between 'Middle' and 'Central' Asia. In the narrow sense, Middle Asia is an area between the Caspian Sea in the west and the Pamir Mountains in the east, the Aral-Irtysh watershed in the north, and the Kopetdag-Hindukush Mountains in the south. At present, administratively, 'Middle' Asia includes the newly independent states of the former USSR: Turkmenistan, Uzbekistan, Tadjikistan, Kyrghyzstan and southern Kazakhstan, which form a separate geopolitical region. These countries have much in common: the physical environment, economic heritage from the former Soviet Union, similar traditions in agriculture and a similar culture and religion (Islam). Many environmental problems have been inherited from the socialist period; some of them have been aggravated recently after obtaining the independence. Among such problems, the main one is desertification expressed in desert pastures degradation due to overgrazing and (in some areas) undergrazing, water erosion in the mountainous areas and piedmonts due to the cutting of trees and shrubs for firewood, soil salinisation as a result of the outdated irrigation techniques applied in the most of cultivated fields, waterlogging of the vast desert areas by drainage and excessive irrigation waters flowing to the adjacent to cultivated fields areas.

One of the most distributed forms of land degradation in the Middle Asia is wind erosion leading to the development of dust storms. Dust storms are one of the most characteristic features of the region. These events begin by the wind velocity 9–14 m/sec and sometimes with a lower velocity of about 6–8 m/sec. The greatest

number of days in a year with dust storms was registered in the following stations: Cheshme – 113 days (1948), Repetek – 106 days (1939), Molla-Kara – 146 days (1938) (all three located in Turkmenistan); Muinak – 121 (1958) (Uzbekistan). Thus, the study of the frequency and spatial distribution of the dust storms is of great importance both for the local communities and policymakers, since they affect not only the agricultural and the other day-to-day activities, but also the health of the people living in the drylands of Middle Asia. In Karakalpakstan (Autonomous Republic in the north of Uzbekistan), the incidence of childhood pneumonia is the highest in the former USSR; in Turkmenistan, 50 per cent of all reported illnesses are the respiratory diseases; lung diseases among children are critical in Kazakhstan (Wiggs et al. 2003). Salt and dust aerosols contain biochemical matter and radioactive nuclides, which develop diseases of the central nervous system, blood, parenchymal organ, cancer, hepatitis etc. (Tolkacheva 2007).

Regular monitoring of dust storms began in this region in the 1930s and was conducted at a large number of observation sites located over the entire territory of Middle Asia. The first studies, which dealt with the analysis of dust storm observations, were published during the 1960s. Romanov (1960) was the first to classify and analyse the frequency and duration of dust storms for the period 1951–1955 at 40 meteorological stations located all over Soviet Middle Asia. In the same work, 17 synoptic processes (large-scale weather patterns), which are significant for dust storm formation, were identified. Orlovsky (1962) analysed the frequency, duration and spatial-temporal distribution of dust storms for Turkmenistan. In the 1970s, a number of researchers (Chirkov 1970; Sapojnikova 1970) concluded that dust storm activities over the arid territories of the USSR are unevenly distributed due to the large variety of soil surfaces, and increase their frequency from the north to the south. Starting in the 1960s, the Middle Asian region began experiencing major land-use and land-cover changes. In Kazakhstan, land degradation began in the 1950s–1960s due to the 'development of virgin lands' policy, the increase in animal numbers and changes in grazing practices (Robinson et al. 2003). The major factor in the land degradation of Turkmenistan and Uzbekistan was the irrigated agriculture expansion programmes, which led to the transformation of vast areas of natural desert pastures into cotton fields (Saiko and Zonn 2000; Wiggs et al. 2003). Quite rapidly, one of the major consequences of the landscape transformations became the increase in dust storm frequencies for a significant part of the territory of Middle Asia.

At the beginning of the irrigation expansion programme, a number of scientists expressed concern about the future of the Aral Sea basin, which experienced a critical reduction of water inflow. In 1960, the water flow from the Amu Darya and Syr Darya Rivers, which represented the major input for the Aral Sea, varied from 56km^3/year to 60km^3/year (Micklin 2014); by the mid-1980s, the water flow had almost ceased, and the water volume discharged was 3.5km^3/year (Saiko and Zonn 2000). By 2007, the water inflow to the Aral Sea was about 5km^3/year–10km^3/year (Dukhovny and Stulina 2011). According to Wiggs et al. (2003), since 1960 the sea level has declined substantially, with a 53 per cent decrease in surface area and a 70 per cent decrease in volume. Two years before the collapse of the Soviet

Union in 1991, the territory was proclaimed an 'ecological disaster zone' (Saiko and Zonn 2000). In 1960, the water volume of the Aral Sea was 1093.0km^3; in 1986, it had decreased by more than half to 448.00km^3, continuing to decrease to 105 km^3 by 2009 (CAWATERinfo). In 1960, the Aral Sea area was 67,500km^2 (Micklin 2010), while, in 2009, the water surface occupied the area of 13,500km^2. In 1987, the fourth biggest inland water body in the world split into two parts: the northern – the Small Aral, and southern – the Big Aral (the latter, by 2009, had separated into deep western and shallow eastern parts). In 2005, a 13-km-long dike was constructed blocking the inflow of Syr Darya River water into the Big Aral Sea in order to preserve, at least, the Small Aral. By 2000, 42,000km^2 of the seabed was exposed resulting in a new desert – the Aralkum. The analysis of recent satellite images provides evidence that in August 2011, the area of the Aral Sea's dried bottom was 57,529km^2. The water bodies' areas were 2,317km^2, 4,411km^2 and 3,243km^2 for the Eastern, Western and Small Aral Seas, respectively (Kozhoridze et al. 2012). As a result of the Aral Sea's level drop, new dry areas became active hotspots of dust storm outbreaks.

The critical changes in the hydro-meteorological regime of the Aral Sea led to the soil degradation and desertification of huge areas. Dust storms occurring in Central Asian natural and anthropogenic deserts transport large amounts of material over long distances and significantly affect agricultural activities and human health. The salt-dust plumes can be up to 800km long, and finer (PM_{10}) particles can travel up to 1,500km away, reaching northern Iran, western Turkmenistan and Tien Shan and Pamir mountains (Semenov 2012). Agricultural activities mainly suffer from the intensification of the soil salinisation process, the salinity increase in surface and ground waters, the degradation of pastoral vegetation and the decrease in agricultural crops.

A powerful source of dust and salt aerosols is the newly dried bottom of the Aral Sea, which, for millions of years, was the receiver of the Aral basin's salts and, for the last three-to-four decades, served as an accumulator of fertilizers, pesticides, herbicides and other chemicals washed from the irrigated massifs of the region (Orlovsky and Orlovsky 2001). Due to the rapid transformation processes in the Aral Sea's exposed bottom, the amount of salts carried for distances of hundreds of kilometres reached 15 million – 75 million ton per year and raised considerable concerns regarding the impact on human health (Saiko and Zonn 2000). By other estimations, the total amount of deflated material from the exposed bottom of the Sea varies from 40 million to 150 million tons (Micklin 2010).

To date there is a significant number of publications dealing with dust storms development in the Middle Asia, most of them in Russian. This manuscript, together with recent papers in the western scientific journals (Orlovsky et al. 2005; Indoitu et al. 2012, 2015) represents the review of the existing Russian/Soviet data on dust storms as well as the original analysis of the temporal and spatial distribution of dust storms and trends of the changes from 1930 to the present. This chapter presents a study based on the statistical analysis of meteorological observations from 400 meteorological stations for the period of 1936–2000 (1936–2005 for some areas) using Geographic Information Systems (GIS) technologies to reveal

the spatial distribution of dust storms over Middle Asia for three time periods. The regression equation predicting the number of hazardous/highly hazardous dust storm days from the total number of dust storm days was elaborated. The data on dust storms that originated from the exposed bottom of the Aral Sea are presented. Dust deposition rates in the selected locations in the Karakum Desert have been monitored and analysed. Main dust emission sites have been identified.

Study area

The research was conducted in three Middle Asian countries: Turkmenistan, Uzbekistan and Kazakhstan (Figure 9.1). The plains of Middle Asia include territories from the shores of the Caspian Sea in the west, to the foothills of Alatau, Tian Shan and Pamir-Alay to the east and southeast. The drylands of Middle Asia cover a total area of 1.89 millionkm^2 with a human population of almost 40 million.

This vast territory is represented by a great variety of litho-edaphic desert types, including sandy, sandy-pebble and pebble, gravel-gypseous and gravel, loamy, loess, clayey and solonchakous deserts. Based on a variety of synoptic processes, rainfall patterns and annual and interannual temperature regimes, the region is divided into two climatic provinces: northern and southern regions. While the

Figure 9.1 Administrative map of Middle Asia (produced from http://www.lib.utexas.edu/maps/commonwealth.html, accessed June 6, 2015). Solid black line shows the approximate border between the northern and southern provinces

northern province is characterised by a cold and dry continental Central Asian type of climate, the southern one is distinguished by a hot and dry Mediterranean climate. Mean annual temperatures range between 5 and 11°C in the northern province, and 13–16.6°C in the southern province. The rainfall amount varies between 80–200mm. Less than 100mm of rainfall is registered annually in the desert regions of Karakum, Kyzylkum, Betpak-Dala and the western Balkash shore.

Materials and methods

To analyse the dust storm climatology for Middle Asia, two databases have been used. The first database is the archival data collections contained in the 'Reference Books of the USSR Climate' (1968a, 1968b, 1968c) and the 'Scientific and Applied References Book of the USSR' (1989a, 1989b, 1989c). The data of monthly 24-year averages of days with dust storms were extracted for almost 400 meteorological stations (MS). The second edition of the climatic reference books called 'Scientific and Applied References Book of the USSR,' which incorporated and summarised the climatic and meteorological data for the years 1936–1980, was published in 1989 (1989a, 1989b, 1989c). The data extracted contained 44 years of monthly averages of days with dust storms, their frequencies and durations for every MS. Data for the period 1980–2005 were compiled through the EU-granted CALTER project from the National State Meteorological Agencies of Turkmenistan, Uzbekistan and Kazakhstan. The dataset contains the total number of days with dust storms for each month/year.

Arc Map software was used as the main tool to analyse the spatial and temporal distributions of the dust storm events. The spatial and temporal distribution maps were constructed based on averages for three periods of time: 1936–1960, 1936–1980 and 1980–2005. The multi-year variability of the dust storm frequencies was analysed using data for 1936–2005. To analyse the dust storms that originated from the Aral Sea's dried bed, we used the archive of NOAA AVHRR (12, 16 and 18) images collected during 2005–2008 at the receiving station established at the Kazakh Space Research Institute in Almaty, Kazakhstan. A total of 42 images were used to identify the locations of dust and salt storm source areas over the 'new' desert, as well as to determine the changes in the transport direction, sizes and structures of the dust plumes.

Dust deposition rates were studied at six locations in Turkmenistan's Karakum Desert in May–October, 2009, using dust traps. Monitoring was carried out using horizontally installed dust traps, which were especially designed to avoid the rebound and fall of dust particles, in May–October, 2009. Each trap consisted of a circular plastic tray of 22.5cm in diameter (area of $0.04m^2$) and 3cm depth. A carpet of artificial grass (with a plastic foil beneath the carpet) was placed in the tray. Fine netting was stretched over the surface of the tray to reduce organic contamination (small insects, bird droppings, small leaves and plant residues, etc.). The plastic tray was attached to a 0.65m-long wooden shoulder and then mounted at a height of 3.0m above the surface on a post at each observation site. The placement at this height followed the international meteorological standards for aeolian

measurements in order to minimise turbulences caused by the boundary layer (e.g. the deposition of large local saltating sand particles from the soil surface) and to prevent human and animal interference. The trap inserts of the two parallel samplers were put in place on the first day of each month and left there until the end of the month. The dust sampler inserts were emptied and changed monthly.

The one sampler (for dust storm events) was kept covered on the upward-facing side during normal meteorological conditions. It was opened at the beginning of a dust storm; after the end of the dust storm the inserts (artificial grass carpet and the underlying plastic foil) were changed. In the case of a dust storm event, the precise time (days, hours and minutes) of the single event were carefully recorded. The plastic bags (with dust probes) were kept in a dark, dry place. The dust was removed from the artificial grass carpet and the plastic foil with a small synthetic brush and was put into little plastic zip-lock bags, sealed and marked with the name of the station, the month/year and the duration of the dust storm in hours/minutes. The dust samples were weighed using a precision scale with an accuracy of 0.0001g, and a total error (due to scale calibration and package weight) of 0.78 per cent.

Results and discussion

Geographic distribution of dust emissions

The vast expanse and diverse desert types across Middle Asia generate dust storms of different frequencies, intensities and durations. The spatial distribution of dust storms for 1936–1960 and 1980–2000 are shown in Figures 9.2 and 9.3. During

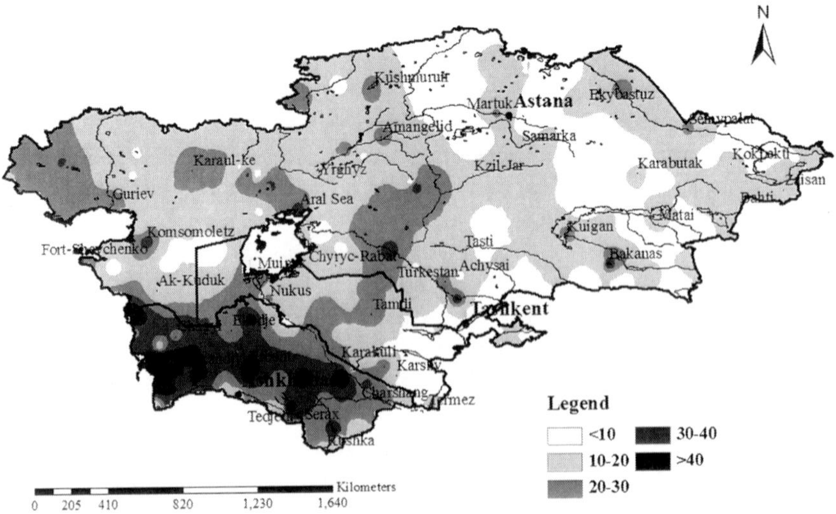

Figure 9.2 Spatial distribution of the annual average frequencies of dust storms in 1936–1960

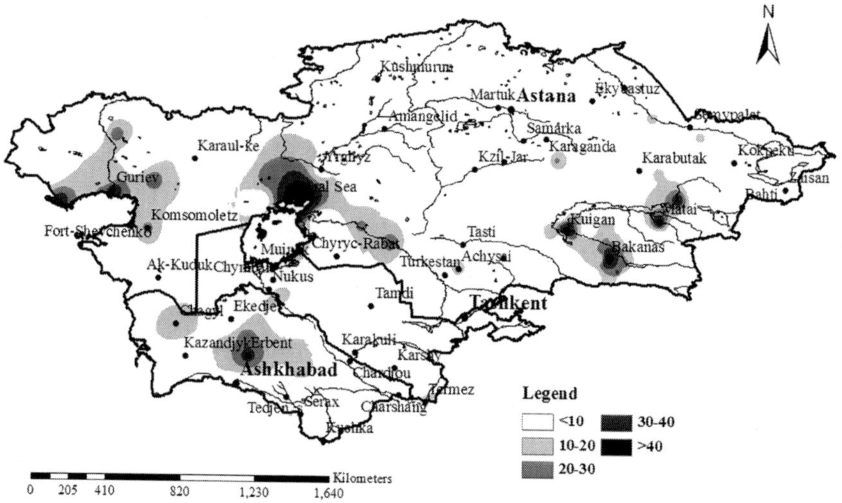

Figure 9.3 Spatial distribution of the annual average frequencies of dust storms in 1980–2005

the 1936–1960 period, Central Asia was affected by a greater number of dust storms compared to the following decades; the region was affected, on average, by 10 to 20 dust storm events/year. The measurements showed that quarter of all meteorological stations (MS) registered more than 40 days of dust storms per year (Figure 9.2).

The largest dust storm source area was the Karakum Desert. The desert, as a whole, was an enlarged source of dust storm frequent outbreaks with more than 30 days of dust storms per year, which appeared as a 'belt' stretching from the west to the east. This 'belt' of high frequency dust storms extended from the Pricaspian sands in the west with the maxima of 81 days/year.

In the northern deserts, the situation was slightly different due to the variety of desert lithological covers. There were two large centres with high dust storm frequency: the North Lowlands of the Caspian Sea and the sandy and clayey Kyzylkum Desert to the east from the Aral Sea. The maximum number of days with dust storms was registered at the salty pans of the northeastern Caspian Sea shore – 38 days/year. Other hotspots of dust storms were the northern areas of the Aral Sea and Kyzylorda (35 and 33 days/year respectively) (see Figure 9.2).

In comparison to previous decades, the years following 1980 showed the most visible evidence of a considerable dust storm frequency decrease during the last century (Figure 9.3). The entire region experienced less than 10 dust storm outbreaks per year. The major hotspots of dust storms diminished considerably. The only significant area to generate dust storms was north of the Aral Sea, where the number of dust storms almost doubled, from less than 40 days in the previous decades, to an average of 64 days with dust storms during 1980–2000.

The increase in dust event frequency, undoubtedly, was connected to the desiccation of the Aral Sea. Due to the absence of any ground observation points, remote sensing remains the only method of monitoring the dust storm events on the Aral Sea's dried seabed.

Dust storm trend analysis

Analysis of meteorological data at a global scale showed that in approximately the last six decades some areas have shown increasing trends of dust storms (Sahel), while others (Mongolia, China, Australia, etc.) have shown declining trends (Goudie and Middleton 1992; Goudie 2008). Interannual dust storm occurrences at a global level showed significant decreasing trends in the second half of the 20th century, particularly in the middle of the 1970s (Ekstrom et al. 2004; Wang et al. 2004; Goudie and Middleton 2006). The decreasing trend of dust storm frequency continued until the end of the last century.

A great number of Middle Asian meteorological stations also showed a considerable decreasing trend of dust storm frequency after the 1970s. As the observations of dust storm frequency began back in 1936, the statistical analysis shows that, at the beginning of the 20th century, the region as a whole was not suffering from a great number of dust storm outbreaks. The middle of the century, 1940–1950s, was characterised by upward trends of dust storm frequency for a large number of stations. The first period of time when records showed a significant decrease in the number of days with dust storms was the mid-1960s (in the central and southern Kyzylkum Desert in Uzbekistan, this decrease was observed even earlier, in the middle of the 1950s). The downward trend of dust storm activity was registered at a large number of observation sites, although more prominently in the northern deserts. Later, from the end of the 1960s to the beginning of the 1970s, the frequency of days with dust storms registered a slight increase for the northern deserts and a more considerable increase for the southern deserts. Concomitant with the global dust storm pattern, Middle Asian dust storm outbreaks showed a significant decrease in dust storm frequency at the end of the 1970s/beginning of the 1980s. This discernible pattern was characteristic for most the stations. The 1970s downward trend of dust storm occurrences continued until the late 1990s. During this period of time, the number of dust storm events reduced by two-to-threefold, a reduction monitored at all the stations. The exceptions were the northeastern, eastern and southwestern areas of the Circum-Aral region, where the dust storm frequency has increased since 1980. A few records show a slight increase in dust storm frequency at the end of the 1990s to the beginning of the 2000s. Since 2000, the increase in dust storm activities has been registered at almost all the meteorological stations in the study area; future observations will show how stable this tendency is.

Figure 9.4 shows the pattern of long-term dust storm frequency variations for four meteorological stations located in different areas over Middle Asia (Kyzylorda: 44°51′N 65°31′E, Tedjen: 35°43’N 61°36’E, Jusaly: 45°29′N 64°04′E and Dashoguz: 41°50′N 59°58′E). The graphs show that during 1936–2005, there were

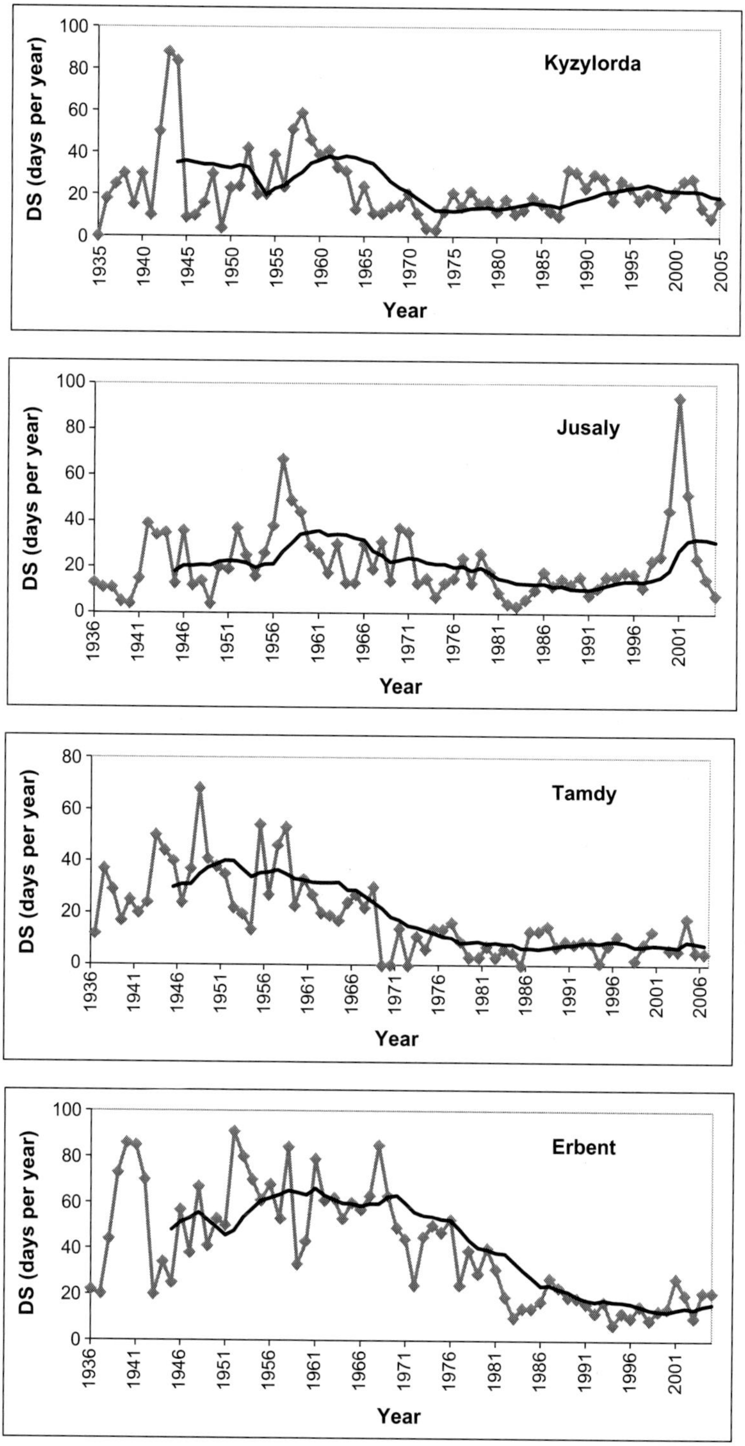

Figure 9.4 Annual dust storm (days) occurrences for the period 1936–2005 (line with markers: total annual days with dust storms; solid line: five-year moving average)

cycles of maxima and minima of dust storm activities, typical of each area in particular. As the observations of dust storm frequency began back in 1936, the statistical analysis showed that at the beginning of the 20th century, nearly the entire region suffered fewer dust storm outbreaks.

Such changes may be explained by the alteration in regional/global atmospheric circulation. Western, northern and northwestern cold intrusions are responsible for the development of 40 per cent of dust storms. In 22 per cent of cases, dust storms occur at the periphery of anticyclones and in 14 per cent of cases at the exits of southern cyclones. Zolotokrylin (1996) notes a 26–29 per cent decrease in the recurrence of these three synoptic processes in 1966–1985 period, relative to the period of 1936–1965, which may explain the decrease in dust storm frequency at that time. At the same time, the average annual wind speeds, measured at several stations in the northwestern Kyzylkum Desert in Kazakhstan also showed a decreasing trend, starting in the mid-1960s and continuing till the early 2000s (Weidel et al. 2004).

The middle of the century – the 1940s – was characterised by upward trends of dust storm frequency for a large number of stations. Later, at the end of the same decade, there was a considerable downward trend present in most of the region. The first time period during which records showed a significant decrease in the number of days with dust storms was the second half of the 1960s. This downward trend of dust storm activity was registered at a large number of observation sites, although more prominently in the northern deserts. Later, at the end of the 1960s and the beginning of the 1970s, the frequency of days with dust storms registered a slight increase for the northern deserts and a more considerable increase for the southern deserts. Concomitant with the global dust storm pattern, Middle Asian dust storm outbreaks showed a significant decrease in dust storm frequency at the end of the 1970s/beginning of the 1980s. This discernible pattern was characteristic for most of the stations. In the 1970s, a significant downward trend of dust storm occurrences continued until the late 1990s. During this period of time, the number of dust storm events reduced by two–threefold, a reduction monitored at all the stations. A few records showed a slight increase in dust storm frequency at the end of the 1990s/beginning of the 2000s (Figure 9.6, Jusaly).

Synoptic processes

Dust storms are predominately controlled by a number of synoptic processes that reflect typical weather over Middle Asia. All types of synoptic processes can be divided into four groups by weather regimes (Table 9.1): cyclonic bursts (types 1–4), cold intrusions (types 5–8, 10, 15), anticyclonic weather (types 9, 9a, 9b, 11–13) and cyclonic activity (type 14). Western intrusions (type 10, see Table 9.1) were identified as the synoptic process that caused most dust storms over Middle Asia (26 per cent).

Figure 9.5 represents a 10-year average of the western intrusion anomaly, which shows clear cyclical occurrences during the last century. The western intrusion pattern showed a negative decline from the mean annual frequency, a decrease

Table 9.1 Characteristic Middle Asian synoptic processes (1936–1990)

Type	*Name of synoptic process*	*Frequency*	*%*
1	South Caspian Cyclone	805	9.2
2	Murgab Cyclone	250	2.8
3	Upper-Amu Darya Cyclone	23	0.3
1+2+3		**1,078**	**12.3**
4	Large-scale warm air mass carrier	81	0.9
5	Northwestern cold intrusion	1,579	18.0
6	Northern cold intrusion	711	8.1
5+6		**2,290**	**26.1**
7	The wave activity	258	2.9
8	Thermal low	301	3.4
9	Southwestern periphery of the anticyclone	901	10.3
9a	Southeastern periphery of the anticyclone	204	2.3
9b	Southern periphery of the anticyclone	663	7.6
9+9a+9b		**1,768**	**20.1**
10	Western intrusion	2,298	26.2
11	Summer thermal low	173	2.0
12	High pressure gradient field	188	2.1
13	Low pressure gradient field	309	3.5
14	Western Cyclone	33	0.4
Total		**8,777**	**100**

of approximately 15 events, from the 1930s until the middle of the 1940s, after which an increasing pattern was observed with the maxima in the 1960s. During the 1960s, until the end of the 1980s, their occurrence increased significantly, with about 10–20 events/year more than the annual average.

The second most-registered synoptic processes (26 per cent) that favoured the occurrence of dust storms were the northern (type 6) and the northwestern cold intrusions (type 5) (Table 9.1). The respective intrusions triggered the worsening of the weather by decreasing the temperature and causing sudden squalls.

The anticyclone intrusions registered their peak (compared to the annual average) during 1940–1960 (about 35 events/year) and reached the maximum during 1950 with 78 outbreaks (Figure 9.5). In the following years, until the middle of the 1970s, the number of southwestern, southeastern and southern peripheries of anticyclones (types 9, 9a and 9b) significantly decreased to almost 10 outbreaks/year and with the minimum during 1969 (four events). From 1976 to 1985, the average of the anticyclone intrusions almost doubled, with the maximum of 58 events during 1977. At the end of the 1980s, the synoptic process pattern showed a major decrease in occurrences.

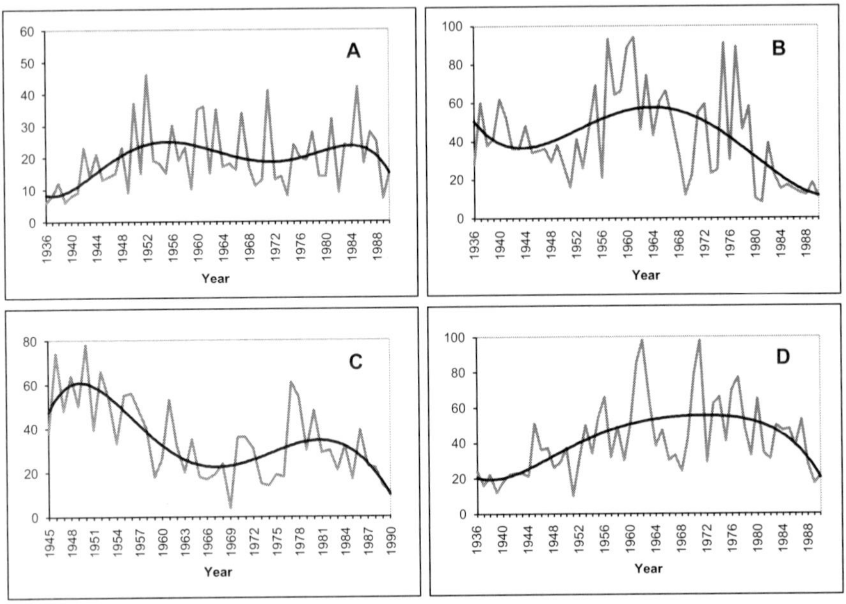

Figure 9.5 Annual average Middle Asian synoptic process anomalies, 1936–1990 (change relative to 1936–1990 average). Axis Y – frequency of synoptic processes types, solid line – 10-years moving average. A: types 1, 2, 3; B: types 5, 6; C: types 9, 9a, 9b; D: type 10

Southern cyclones are another group of synoptic processes that influence dust storm outbreaks (Table 9.1). The mean annual frequency of southern cyclone (types 1–3) occurrences was 20 events per year. From the beginning of record-keeping in 1936, the frequency of southern cyclones gradually increased, reaching the maximum during the 1960s.

A statistical analysis of the synoptic processes' five-year averages during 1936–1990 showed a considerable decline in the frequencies of the four types responsible for dust storm occurrences during the last two decades of the monitoring period. The major decrease in synoptic process frequencies was observed during 1985–1990. This phenomenon was, most likely, the main cause of the dust storm decrease in Middle Asia in the second half of the 20th century.

Severe dust storms

The study of the spatial distribution of hazardous and highly hazardous dust storms by monitoring the number of days of such events in Middle Asia is very important for understanding the dust storm phenomenon. Semenov and Tulina (1978) provided the most credible data on hazardous and highly hazardous dust storms. This paper defines hazardous dust storms as events with the following characteristics: duration of 3–12 hrs, a wind speed of 10–14m/s, and a meteorological visibility

range of 500–1000m. According to Semenov and Tulina (1978), highly hazardous dust storms have the following parameters: a minimal duration of 12 hrs, a minimal wind speed of 15m/s and a maximal visibility range of 50m (irrespective of event duration and wind speed).

The spatial and temporal distributions of dust storms in Kazakhstan were studied by daily meteorological observations at 144 stations during 1936–1972. Days of hazardous and highly hazardous dust storms were calculated on the basis of the abovementioned criteria, and they correlated well with the total number of dust storm days. Data on Uzbekistan and Turkmenistan were collected at 29 weather stations for the same period.

The regression equation predicting the number of hazardous/highly hazardous dust storm days from the total number of dust storm days is as follows:

$$Y=0.4004x-1.9836;\ R2=0.827 \quad (1)$$

where Y is the number of days of hazardous and highly hazardous dust storms, and x is the total number of dust storm days (Orlovsky et al. 2013). This equation allows us to calculate the number of days of hazardous and highly hazardous dust storms in Central Asia. Thus, it is possible to assess the spatial distribution of the days of such events in the studied regions (Figure 9.6).

Large source areas swept by hazardous and highly hazardous dust storms are located in the northwestern part of the studied region, mainly in the Ili River Valley, the Karakum Desert and the central part of the Kyzylkum Desert. The

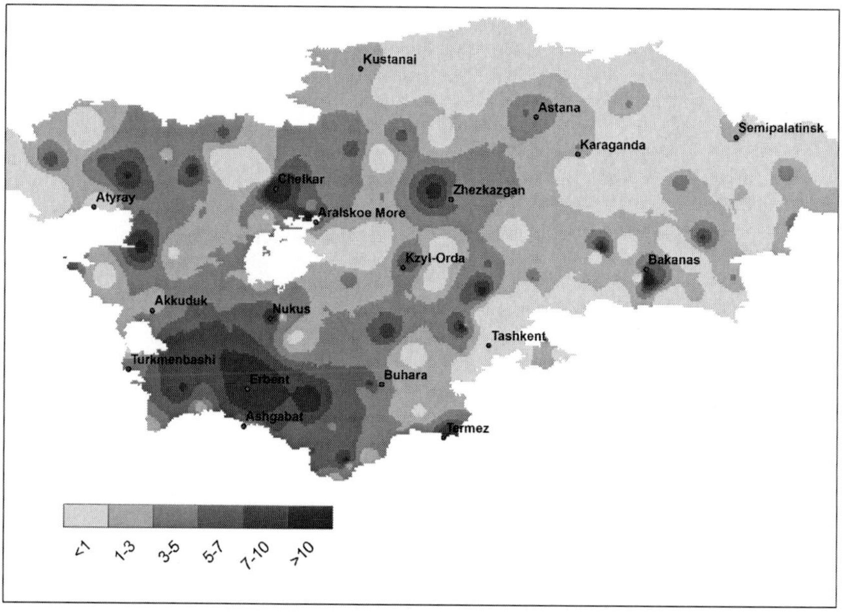

Figure 9.6 Spatial distribution of severe and very severe dust storms in Middle Asia.

highest frequencies of hazardous and highly hazardous dust storms are monitored by weather stations in Central Karakum Desert, Central Kazakhstan, northern Aral region and northeastern Caspian shore.

Amount of dust deposits

There has been much speculation on the amount of dust emissions/deposits in Middle Asia and a very limited amount of measured data. The high values for dust deposits were registered on the dried bottom of the Aral Sea – 2,400–2,700kg/ha per year (Tolkacheva 2000). Farther from the dried bottom zone, the amount of dry atmospheric deposits decreases, and at a distance of 600 km from the dried strip, they amount to 100–150kg/ha per year. The nature of the dry atmospheric deposits varies within the year and has a seasonal character. The deposition of the largest portion of aerosols occurs during the summer dust storms. Thus, in the Uyaly area (45⁰15'N, 61⁰14'E, located 15km from the Sea for the period of cited observations, about 90km at present), the deposition of sand and dust during a single dust storm event in July may vary from 450 to 1,800kg/ha (Tolkacheva 2000).

Observations in the 1980–1990s on salty dust transfer in 43 points of the Aral region revealed that the deposition of salty dust aerosols depended significantly on the topography of the area, prevailing winds, vegetation cover, soil moisture etc., and reached the value of 1,500–6,000 t/ha, where the portion of soluble salts were 170–800kg/ha (Razakov and Kosnazarov 1996). According to the authors, the volume of dust/salt deposits on the dried bottom reached 1,600kg/ha at that time.

Similar data for aerosol deposits were obtained at 16 observation points in Karakalpakstan for the period of May–October 2000 by Wiggs et al. (2003). Deposition rates were highly variable throughout this period, with values ranging from less than 500kg/ha at Nukus (42⁰28'N, 59⁰36'E, 200km from the Sea) to nearly 2,500kg/ha at Kyzyl-Djar (43⁰34'N, 59⁰06'E), previously located on the Aral Sea shore and currently about 55km away from it. Agricultural sites experienced variable rates (170–270kg/ha) throughout May–August before declining to 120kg/ha in October. The agricultural sites adjacent to the former shore of the Aral Sea showed rates of 560kg/ha in July and August before declining to 380kg/ha in September. These results indicate high levels of dust deposition throughout the southern Aral region in the summer months, but with particularly high rates of deposition in the north of the region close to the shore of the former Aral Sea in July, August and September.

The Karakum Desert is the most active source of dust emissions in Middle Asia. In the course of the CALTER project funded by the European Commission, dust samples were collected at six sites located in the different litho-edaphic types of the deserts in order to assess the amount of dust deposits. The monitoring site locations were as follows: 1. ***Kunya-Urgench*** (42°19'25.3"N and 59°10'1"E); 2. ***Shakhsenem*** (41°44'38"N and 58°43'14"E); 3. ***Darvaza*** (40°11'13.02"N and 58°25'2.05"E); 4. ***Bokhordok*** (38°45'32.3"N and 58°29'39.85"E); 5. ***Chagyl*** (40°47'0"N and 55°23'0"E); 6. ***Repetek*** (38°33'53.41"N and 63°10'36.13"E). Table 9.2 shows the amount of dry deposits (total amount including deposits during the dust storms).

Table 9.2 Dust deposits (kg/ha) in Turkmenistan during May–October 2009

Month	*Bokhordok*	*Chagyl*	*Darvaza*	*Kunya-Urgench*	*Repetek*	*Shakhsenem*
May	2,374.9	34.9	283.8	251.0	163.7	119.6
June	1,742.1	483.7	113.9	214.2	267.0	95.1
July	1,142.7	82.6	90.9	176.0	313.3	42.1
August	567.3	57.9	119.3	73.1	480.6	114.2
September	758.8	21.6	123.7	97.2	239.5	71.2
October	186.0	279.4	208.3	117.5	123.4	73.6
Total	6,771.8	959.8	939.9	929.0	1587.5	515.8

During the monitoring period, the highest dust deposition rate during dust storms (c.a. 4500kg/ha for observation period) occurred in Bokhordok (closure area of clayey desert of the northern Afghan rivers and sandy deposits of the Karakum suite) – the station with the highest frequency and duration of dust events. The station with the second highest dust deposition rate of 950kg/ha was Repetek, located in the juncture of the sandy deserts of Central and Southeastern Karakum. The other four stations had much smaller dust deposition rates: in Kunya-Urgench (clayey desert in the ancient Amu Darya River Delta newly developed agricultural area), 340kg/ha; in Darvaza (sandy desert at the border of Central and Transunguz Karakum), 260kg/ha; in Shakhsenem (area of newly developed irrigated lands in northern Turkmenistan), 160kg/ha; in Chagyl (sandy massive in the northwestern Turkmenistan), 100kg/ha.

In regular weather conditions, the distribution of dust deposition differed from that during dust storms. The highest dust deposition rate in regular weather (c.a. 2300kg/ha) was also registered in Bokhordok. The station with the second highest dust deposition rate of 860kg/ha was Chagyl – the station with the lowest frequency and duration of dust events. In Kunya-Urgench, Repetek and Darvaza the regular weather dust deposition rates varied between 620 and 680kg/ha; the lowest dust deposition rate was measured in Shakhsenem, of 370kg/ha. Only in Bokhordok and Repetek (the two stations with the highest frequency and duration of dust events) was the dust deposition rate during dust storms higher than that in regular weather. In the other stations, the total amount of dust deposited in regular weather was larger than that during dust storms.

During the monitoring period, the highest frequency and duration of dust storms, along with the highest dust deposition rates, were registered in Bokhordok. We suggest that this was mainly induced by natural factors: the location at the northern part of the so-called takyr-ridge complex, and the presence of the vast sandy massifs. The long-term monitoring of dust storms clearly identifies the aforementioned area as a region of dust storms with high frequencies and durations (Orlovsky et al. 2005). In addition, a high concentration of livestock around this settlement has caused significant overgrazing of the sandy pastures and a reduction of the

soil fixedness, which has eventually led to an increase in dust emission from the underlying surface.

Dust storms originating from the dried bottom of the Aral Sea

In the 1990s, the western world became aware of the ecological disaster of what was the fourth largest lake in the world, the Aral Sea. The drastic desiccation of the Aral Sea led to the intensive development of desertification processes in the region and the occurrence of a new desert, the Aralkum, on the dried sea bottom. Estimates of the total deflated material from the dried bottom of the Aral Sea, which were made in the late 1980s, ranged from 13 million to as high as 231 million metric tons/year (Glazovskiy 1990). In addition, in the Southern Aral Sea, there are more than 1 million ha of takyr and takyr-like soils, which constitute the highest potential for severe dust storms over the area (Singer et al. 2003).

According to the satellite imaging, the dust storm plumes can stretch for lengths of 150–300km, with a maximum of 500km (May 6, 1979) (Grigoryev and Jogova 1992). The first estimations of dust volumes transported during the dust storms in 1975 were, on average, 45 million t/year (Grigoryev and Lipatov 1983), while in 1990, this figure doubled (90 million t/year) (Grigoryev and Jogova 1992). In addition to the fine grains carried from the sandy and solonchak surfaces, the dust plume loads contain salts, such as sulphates and chlorides, which are deposited over an area of 250,000km^2 (Kondratyev et al. 1985).

The monitoring of dust storm outbreaks, over the new-developing desert on the dry bottom of the Aral Sea, revealed that the temporal distribution of the dust storm events increased significantly. The long-term remote sensing data analysis for the period 1975–1982 showed a total of 37 dust storm events; from 1985 to 1990, 33 dust storm outbreaks were registered (Grigoryev and Jogova 1992). From 1990–2002, the measurements executed at nine meteorological stations in the Circum-Aral region registered 1082 days with dust storms of different intensities and durations (Shardakova and Usmanova 2006). From 2005–2008, a total of 30 dust events of different durations, sizes and structures, originating from the Aral Sea's dried bottom, were registered by NOAA images, from which about 70 per cent occurred on the 'old' dried seabed.

Transfers of salt and dust are usually directed to the south-southwest, towards the Amu Darya River Delta and the Ustyurt Plateau, but in some cases, they stretch to the east-southeast towards the Syr Darya River Delta (Grigoryev and Lipatov 1982). An analysis of space images for 1975–1981 revealed that in 60 per cent of all cases, the salt and dust blowouts moved to the southwest to oases in the Amu Darya Delta, in 25 per cent of events to the west to the Ustyurt Plateau, and in the rest of the cases, to the south and southeast (Grigoryev and Lipatov 1983). The images from 1975 showed that the dust plume hardly ever reached the western shore of the sea (Grigoryev and Lipatov 1983), and mainly precipitated on the water surface (Kondratyev et al. 2002). However, during the dust outbreaks of

1979, most of the dust plumes reached the adjacent shore, often on the south and southwestern shores (Kondratyev et al. 2002).

The area of salt and dust deflation comes to 20,000–30,000km^2; raised and transported salt and dust affects the surrounding territory, more than 500,000km^2. It should be recognised that the satellite images only register cases with very strong atmospheric turbidity, so in reality, dust and salt could be transported to distances much further than can be observed from space. Based on this assumption, Grigoryev and Lipatov (1983) supposed that salt from the Aral region reaches the Caspian shore and spreads all over the Aral-Caspian lowlands. From 2005–2008, the length of the dust plumes reached values from 150 to more than 600km. The areas covered by the dust plumes also varied from a few thousand km^2 to a few hundred thousands of km^2. Air masses intruding from the east (57.1 per cent), northeast (11.9 per cent) and southeast (9.5 per cent) were observed to create the largest dust storm plumes.

Sources of the salt and dust emission

Weakly stabilised and barren sands, secondary salinised irrigated lands and the dried bottom of the Aral Sea serve as sources of atmospheric pollution in Middle Asia. Relatively recent publications (Singer et al. 2003) show that takyrs and takyr-like soils constitute the surfaces with the highest potential for being the source for the area's dust storms. Takyrs and takyr-like soils occupy over 120,000km^2 in Middle Asia (Mamedov 2014), and 6,800km^2 in the dried bottom of the Aral Sea (Kozhoridze et al. 2012). Second to the takyr soils, the solonchaks and solonchak-like soils also contribute highly saline dust. Solonchaks (salty pans) are typical for the Middle Asian deserts; they occupy an area of more than 150,000km^2 (Pankova et al. 1996). Solonchaks occupy 22,500km^2 of the 57,500km^2 (data for 2011) of the Aral Sea's exposed bottom (Indoitu et al. 2012). For the 'old dry zone,' a relatively low soil salinisation rate is characteristic, with sulphates, hydrocarbonates and calcium. In the surface layer of the soils of the 'newly dried zone,' the sulphates, chlorides and sodium dominate (Tolkacheva 2000).

The salty crust in the sandy and sandy-loam soils from the dried bottom of the Aral Sea is being destroyed and removed by even light winds with a velocity of 2.5–5.0ms^{-1}. At the same time, experiments with soil samples from the dried bottom demonstrated that the materials from the sites with the highest dust generation also had the lowest salt contents (Singer et al. 2003). Large areas in the dried bottom of the Aral Sea are composed of sand dunes, and the winds are loaded with saltating sand particles. The area occupied by sandy fields, in combination with takyr-like soils, was almost 11,200km^2 of the exposed bottom in 2009, and the area under sandy massifs with solonchaks increased from 1,000km^2 to 17,600km^2 from 1989–2009.

The analysis of the active dust emission sites in the Aral Sea's dried bottom revealed that newly formed sandy massifs, combined with dry puffy solonchaks with easily destructible top surfaces, and takyrs represent the main dust source

from this location. The active emission site consists of sands (75 per cent), solonchaks (17 per cent) and takyrs/takyr-like soils (8 per cent).

Concluding remarks

An analysis of data on dust storms for the last century demonstrated the clear downward trend of dust storm frequency in Middle Asia. The significant decreasing trend was most obvious over the Karakum Desert where dust storm occurrences reduced from an average of more than 30 days per year to less than 20 days per year. An analysis of the spatial distribution of dust storms suggested four main dust source areas, which, during the study period, underwent changes, such as shifting and shrinking their areas. The main spatial changes were obvious in the northern Caspian deserts, where dust storm occurrence areas shrunk significantly and also underwent a few hundred kilometres shift to the east. The Karakum and Kyzylkum Deserts, as well the south Balkhash Lake area, experienced an important surface reduction of the major source areas of dust storm activities. As all the hotspots of the highest dust storm activities showed a decreasing trend of dust emissions, the new Aralkum Desert became very active in the last two decades of the 20th century.

During the last three decades, various scientists have observed a trend characterised by a significant decrease in dust storm frequency over the major part of Central Asia. The increasingly frequent sources of hazardous and highly hazardous dust storms coincide with the increasingly frequent sources of dust storms whose total duration of dust storm days exceeds 20 days. Large areas of hazardous and highly hazardous dust storms are located in the northwestern part of the analysed region, mainly in the Ili River Valley, as well as in the central parts of the Karakum and Kyzylkum Deserts. The highest frequency of hazardous and highly hazardous dust storms has been monitored by weather stations in Central Karakum Desert, Central Kazakhstan, northern Aral region and northeastern Caspian shore. The highest dust deposition rates during the summer 2009 were measured in the sandy areas of the Karakum Desert, between 1,600kg/ha and up to 6,800kg/ha. The dust deposition rates at other locations were about 950kg/ha.

The analysis of the NOAA AVHRR images showed that the number of dust storm events originating from the Aral Sea's dried bottom is increasing, and dust particles have been carried out both from the 'older' and from the 'newer' dried bottom of the Aral Sea. In most cases (70 per cent), the dust plumes originated from the eastern terraces dried before 1999. From 2005–2008, a large number of dust plumes had their origins on the surfaces dried before 1990, close to the 1960 water line.

An image analysis for 2005–2008 showed that the main directions of dust plumes' flow changed compared to the earlier observations. From a total of 30 dust storm events extracted from the available NOAA AVHRR data, a small amount of the dust storm events (10 per cent), originally from the eastern shore, were directed toward the east. This direction was not registered in the previous observations. It was also observed that during these four years, 55 per cent of

the dust events were directed toward the west, 30 per cent more than what was observed in the 1970s.

Due to the absence of meteorological and monitoring stations on the dried bottom of the Aral Sea, the only reliable source of information on the number of dust storm events, emission sites and the direction of dust transport is remote sensing information. Dust storms affect all the aspects of the people's life in the region. The blowing dust from the exposed bottom of the Aral Sea, contaminated with agricultural chemicals, became a serious public health hazard. In spite of the decreasing trend in the dust storms frequencies from 1980–2000, in the last decade, there is evidence for a new trend in dust storms activiation. It is possibly connected with change in the global atmospheric circulation – from zonal to meridional, as well as changes in the regional climate – redistribution of precipitation during the year, increase in the climate continentality and temperature extremes. Loss of the Aral Sea is leading to shorter, drier summers, longer, colder winters, and decreased precipitation that contributes to development of the dust storms (Glantz 1999). Reduced growing seasons lead to decreased agricultural productivity and to additional virgin areas being converted into agricultural production.

Anthropogenic factors play significant role in development of the wind erosion processes – governmental policies on increasing the livestock population along with absence of water delivery system to the distant pastures and decrease in number of existing water wells. The switch from cotton cultivation to grain production, which is harvested in June/July, means the land is exposed to wind erosion during summer and with ensuing challenges. Dust emission from the Aral Sea exposed bottom affects the local and regional population – dust aerosols don't recognise the state borders as they can flow large distances of hundreds of kilometres. Dust that originated in the lowlands of Middle Asia is deposited on the Pamir and Tien Shan mountains which contributes to the glacial melt of these main river sources (Aizen et al. 2004; Wu et al. 2009). The salty dust emitted from the sea bottom settles onto fields, thus aggravating the soil salinisation processes. Additional volumes of water are required to flush the croplands, which leads in its turn to further deterioration river water downstream, waterlogging etc. Thus, poorly thought-out decisions in the relatively recent past, such as over-regulation of the rivers' flow, monoculture crop cultivation, irrational and outdated irrigation practices and widespread drainage development instead of cultivation of smaller areas with modern irrigation techniques etc., evoked a vicious circle of cause-effect relationship. To solve the complex existing environmental problems in post-Soviet Middle Asia requires considerable political courage, significant financial investment and enormous efforts directed to educational activities of the local farmers and decision makers.

References

Aizen, V., Aizen, E., Mclack, J., Kreutz, K. and Cecil, L. (2004). Association between atmospheric circulation patterns and firn-ice core records from Ilnichek glacierized área, Central Tien-Shan, Asia. *Journal of Geophysical Research*, 109(8), 1–18. doi:10.1029/2003JD003894

Chirkov, Y.A. (1970). Repetition of dust storms over the USSR territories and the possibility of their prediction. *Proceedings of Hydro-Meteorological Center of USSR*, 69, 109–119 (in Russian).

Dukhovny, V.A. and Stulina, B.A. (2011). Water and food security in Central Asia. In: Madramootoo, C. and Dukhovnu, V. (Eds.) *Water and Food Security in Central Asia.* NATO science for peace and security series C: environmental security. Springer. 1–23

Ekstrom, M., McTainsh, G.H. and Chappell, A. (2004). Australian dust storms: Temporal trends and relationships with synoptic pressure distributions (1960–99). *International Journal of Climatology*, 24, 1581–1599.

Glantz, M.H., (Ed.) (1999). *Creeping Environmental and Sustainable Development in the Aral Sea Basin.* Cambridge University Press, Cambridge, UK.

Glazovskiy, N.F. (1990). *The Aral Crisis: Causative Factors and Means of Solution.* Nauka, Moscow

Goudie, A. (2008). Dust storms: Recent developments. *Journal of Environmental Management*, 90(1), 89–94.

Goudie, A. and Middleton, N. (1992). The changing frequency of dust storms through time. *Climatic Change*, 20, 192–225.

Goudie, A.S. and Middleton, N.J. (2006). Changing frequencies of dust storms. In Czeschlik, D. (ed.) *Desert Dust in the Global System.* Springer, Heidelberg, Germany. 167–192.

Grigoryev, A.A. and Lipatov, V.B. (1982). Dynamics and sources of dust storms in the Aral Sea region by satellite observations. *Proceedings of the academy of science of the USSR*, Geographical series, 4. 73–77. (In Russian)

Grigoryev, A. and Lipatov, V. (1983). Distribution of dust pollution in the Circum-Aral region by space monitoring. *Proceeding of the Academy of Sciences of the USSR, Geographical Series*, 4, 73–77.

Grigoryev, A.A. and Jogova, M.L. (1992). Strong dust blowouts in Aral region in 1985–1990. *Proceedings of the Russian Academy of Sciences*, 324(3), 672–675.

Indoitu, R., Kozhoridze, G., Batyrbaeva, M., Vitkovskaya, I., Orlovsky, N., Blumberg, D. and Orlovsky, L. (2015). Dust emission and environmental changes in the dried bottom of the Aral Sea. *Aeolian Research*, 17, 101–115. doi:10.1016/j.aeolia.2015.02.004

Indoitu, R., Orlovsky, L. and Orlovsky, N. (2012). Dust storms in Central Asia: Spatial and temporal variations. *Journal of Arid Environments*, 85(10), 62–70.

Kondratyev, K., Grigoryev, A., Zhvalev, V. and Melentyev, V. (1985). An integrated study of dust storms in the Aral region. *Meteorology and Hydrology*, 4, 32–38.

Kondratyev, K., Krapivin, V.F. and Phillips, G.W. (2002). *Global Environmental Change: Modelling and Monitoring*. Springer, Heidelberg.

Kozhoridze, G., Orlovsky, L. and Orlovsky, N. (2012). Monitoring land cover dynamics in the Aral Sea region by remote sensing. Proc. SPIE 8538, *Earth Resources and Environmental Remote Sensing/GIS Applications III*, 85381V (October 25, 2012). doi:10.1117/12.972306

Mamedov, B. (2014). Takyr wáter-Harvesitng Systems of Central Asia, Their Protection and Rational Use. DPhil thesis, National Institute of the Deserts, Flora and Fauna, Ashgabat, Turkmenistan.

Micklin, P. (2010). The past, present, and future Aral Sea. *Lakes and Reservoirs: Research and Management*, 15(3), 193–213.

Micklin P. (2014). Introduction to the Aral Sea and its region. In: Micklin, P., Aladin, A. and Plotnikov, I. (Eds.). The *Aral Sea: The Devastation and Partial Rehabilitation of a Great Lake*. Springer, Heidelberg. 15–40.

Orlovsky, L. and Orlovsky, N. (2001). White sandstorms in Central Asia. In Youlin, Y., Squires, V. and Qi, L. (eds.), *Global Alarm: Dust and Sand Storms from the World's Drylands*. N Asia Regional Coordinating Unit, Bangkok. 169–201.

Orlovsky, L., Orlovsky, N. and Durdyev, A. (2005). Dust storms in Turkmenistan. *Journal of Arid Environments*, 60(1), 83–97.

Orlovsky, N. (1962). Some data on dust storms in Turkmenistan. *Proceedings of the Ashgabat Hydrometeorological Observatory*, 3, 17–42 (in Russian).

Orlovsky, N., Orlovsky, L. and Indoitu, R. (2013). Severe dust storms in Central Asia. *Arid Ecosystems*, 3(4), 227–234.

Pankova, E. I., Aidarov, I. P., Yamnova, I. A., Novikova, A. F., and Blogovolin, N. S. (1996). *Natural and Anthropogenic Soil Salination of the Aral Sea Basin (Geography, Genesis, Evolution)*. Dokuchaev Soil Institut, Moscow. (In Russian)

Razakov, R. and Kosnazarov, K. (1996). Dust and salt transfer from the exposed bed of the Aral Sea and measures to decrease its environmental impact. In Micklin, P. and Williams, W. (eds.), *The Aral Sea Basin*. Nato Asi Series 2, Environment 12. Springer Verlag, Berlin. 95–102.

Reference Books of the USSR Climate. (1968a). Kazakh SSR. Hydrometeoizdat, Leningrad, Issue 18, Part III. Wind (in Russian).

Reference Books of the USSR Climate. (1968b). Uzbek SSR. Hydrometeoizdat, Leningrad, Issue 19, Part III. Wind (in Russian).

Reference Books of the USSR Climate. (1968c). Turkmen SSR. Hydrometeoizdat, Leningrad, Issue 30, Part III. Wind (in Russian).

Robinson, S., Milner-Gulland, E. J. and Alimaev, I. (2003). Rangeland degradation in Kazakhstan during the Soviet era: Re-examining the evidences. *Journal of Arid Environments*, 53, 419–439.

Romanov, N. (1960). *Dust Storms of Middle Asia*. Samarkand State University, Tashkent (in Russian).

Saiko, T. and Zonn, I. (2000). Irrigation expansion and dynamics of desertification in the Circum-Aral region of Central Asia. *Applied Geography*, 20, 349–367.

Sapojnikova, S. (1970). Map of dust storm frequency over the southern territories of USSR and their adjoining territories. *Proceedings of Kazakh Academy of Sciences*, 65, 61–68 (in Russian).

Scientific and Applied Reference Book of the USSR. (1989a). Series 3, Multi-Year Data, Parts 1–6, Issue 18, Kazakh SSR, Book 2. Hydrometeoizdat, Leningrad (in Russian).

Scientific and Applied Reference Book of the USSR. (1989b). Series 3, Multi-Year Data, Parts 1–6, Issue 19, Uzbek SSR, Book 2. Hydrometeoizdat, Leningrad (in Russian).

Scientific and Applied Reference Book of the USSR. (1989c). Series 3, Multi-Year Data, Parts 1–6, Issue 30, Turkmen SSR. Hydrometeoizdat, Leningrad (in Russian).

Semenov, O. E. (2012). Dust storms and sandstorms and aerosol long-distance transport. In Breckle, S.-W., Wucherer, W., Dimeyeva, L. A. and Ogar, N. P. (eds.), *Aralkum – A Man-Made Desert: The Desiccated Floor of the Aral Sea (Central Asia)*. Springer-Verlag, Berlin. 73–82.

Semenov, O. E. and Tulina, L. P. (1978). Spatial and temporal distribution of dangerous and especially dangerous dust storms on the Kazakhstan. *Proceeding of KazNIGMI*, 71, 62–77 (in Russian).

Shardakova, L. Yu and Usmanova, L. V. (2006). Analysis of dust storms in Circum-Aral region. *Problems of Desert Development*, (3), 30–34 (in Russian).

Singer, A., Zobeck, T., Poberezhsky, L. and Argaman, E. (2003). The PM10 and PM2.5 dust generation potential of soils/sediments in the Southern Aral Sea Basin, Uzbekistan. *Journal of Arid Environments*, 54, 705–728.

Tolkacheva, G.A. (2000). *Guidance for Monitoring of Dry Atmospheric Deposits in the Central Asian Region.* SANIGMI, Tashkent.

Tolkacheva, G.A. (2007). *Long Term Cological Research Program for Monitoring Aeolian Soil Erosion in Central Asia.* Hydro-Meteorological Institute, Tashkent, Uzbekistan.

Wang, S., Wang, J., Zhou, Z. and Shang, K. (2004). Regional characteristics of three kinds of dust storms events in China. *Atmospheric Environment*, 39, 509–520.

Weidel, H., Roubine, V., Van Daijk, A., Budnikova, T.A., and Dukhovniy, V.A., (2004). Economical Assessment of Joint and Local Measures for the Reduction of Socio-Economical Damage in the Coastal Zone of Aral Sea. Final report INTAS-2001-01059 project.

Wiggs, G.F., O'Hara, S.L., Wegerdt, J., Van Der Meer, J., Small, I. and Hubbard, R. (2003). The dynamics and characteristics of aeolian dust in dryland Central Asia: Possible impacts on human exposure and respiratory health in the Aral Sea basin. *The Geographical Journal*, 169, 142–157.

Wu, G., Xu, B., Zhang, Ch., Gao, Sh. and Yao, T. (2009). Geochemistry of dust aerosol over the Eastern Pamirs. *Geochimica et Cosmochimica Acta*, 73(4), 977–989.

Zolotokrylin, A.N. (1996). Dust storms over Turanian Lowland. *Bulletin of the Russian Academy of Sciences, Geography Series*, 6, 48–54 (in Russian).

10 Climate hazards in Asian drylands

Troy Sternberg

The disruptive nature of climate hazards place Asian drylands at the forefront of climate-human interaction. The 1.4 billion inhabitants of Asian deserts and semi-deserts are vulnerable to one of the most challenging sets of natural hazards of any environment with primary risks driven by climate: scarce and unreliable precipitation, extreme temperature, weather events, drought and dust storms (White and Nackoney 2003; IPCC 2007; Middleton et al. 2011; Middleton and Sternberg 2013). Hazard implications can be extreme – drought was a causal factor of the Syrian Civil War and Arab uprisings (Sternberg 2012; Kelley et al. 2015; Werrell et al. 2015). Floods resulted in great damage in Pakistan (2010), ice storms in China (2008) affected 77 million people whilst extreme temperatures in India (heat, 2015) and Mongolia (cold, 2010) led to increased mortality and major economic losses (EmDat 2015). As home to marginal environments and Asia's poorest communities, arid regions are at the centre of the continent's climate-human predicament as deserts are particularly vulnerable to changing climate, expanding populations, degradation and social challenges (Safriel et al. 2005). This chapter takes a continent-wide perspective to examine unpredictable climate hazard risk across Asia's extensive and diverse dryland regions.

Today 87 per cent of Asia's disasters are climate related, making hazard knowledge critical to human welfare and environmental viability (ISDR 2009; Guha-Sapir et al. 2012). Recognised for their physical damage and economic cost, hazards also have social, political and environmental implications. Recent events in the Middle East exemplify how hazards magnify water and food scarcity, human well-being, migration, state stability and conflict across transboundary landscapes. Past state and international hazard engagement has focused on a country scale and crisis response, yet events, framed by geography rather than governments, have potential global implications. International attention focuses on global frameworks such as the International Strategy for Disaster Reduction (ISDR) and the Hyogo Framework for Action yet mitigation remains a global challenge (ISDR 2011; Sternberg and Batbuyan 2013). This chapter presents human-hazard analysis that stresses desert conditions and commonalities across Asia.

Asian drylands are critical environments because of their extent, exposure and centrality to the continent. Vast dryland expanses cover >18.4 million km^2 (an area larger than South America) and arc from the Mediterranean to the Pacific to

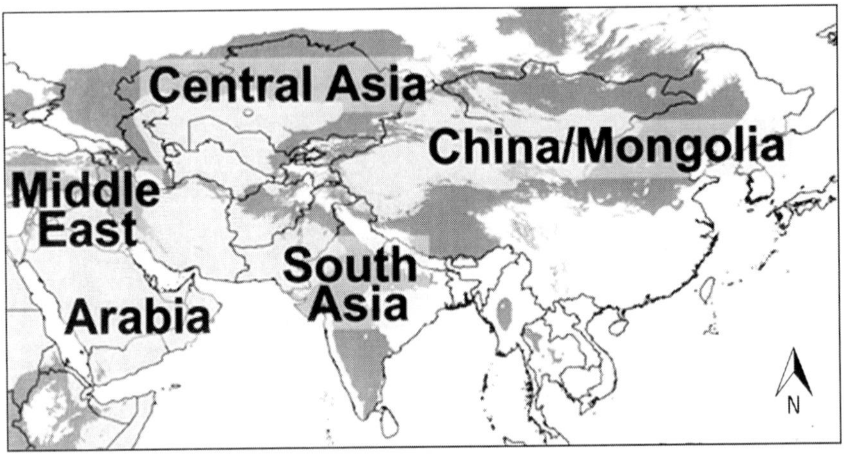

Figure 10.1 Asian drylands, including the West (Middle East, Arabia), South, Central and East Asia. Gray = arid (<250mm annual precipitation), black = semi-arid (<500mm annual precipitation) (Meigs 1953). Dryland percentage varies from ~99 per cent in most of West Asia, to 80 per cent in Pakistan and Mongolia, 69 per cent in India and 52 per cent in China.

encompass a great range of physical, social and political contexts (see Figure 10.1) (White and Nackoney 2003; FAO 2009). First promulgated by Meigs (1953), drylands cover >40 per cent of the globe and are home to ~2 billion people; Asia's arid and semi-arid zones are home to 70 per cent of the world's dryland residents and represent a third of global dry regions (FAO 2009; Sternberg et al. 2015). Expanding populations and rapid economic growth challenges landscapes and livelihoods, placing water security, land use and human development at the forefront of human-hazard interaction in Asian drylands (Scheffran et al. 2011). These conditions focus research on how societies in marginal, at-risk environments engage with the challenges hazards present (Safriel et al. 2005). Surprisingly, despite similar physical factors, Asian drylands are seldom studied as an integrated biome though their risks differ significantly from temperate and tropical zones (Middleton and Sternberg 2013). The expert IPCC (2007) calls for 'innovative (academic) research on the response of natural systems to multiple stresses.' The IPCC further identifies East, Central, South and West Asia as landscapes where food, water, land degradation and biodiversity are considered 'highly vulnerable' to warming temperatures and climate change.

Asian drylands

Asian drylands are comprised of a subtropical high pressure belt that extends from Arabia to South Asia and high latitude, cold deserts that stretch from the Aral Sea to Mongolia and China. Defined by precipitation levels and aridity indices, arid zones (<250mm annual precipitation) and semi-arid regions (<500mm yearly

precipitation) have high evaporation rates due to solar radiation and wind that rapidly removes moisture from the ground surface (Meigs 1953; Sternberg et al. 2015). Regional classification of arid, semi-arid and dry subhumid zones reflects the ratio of precipitation to potential evapotranspiration (P/PET), with aridity signifying a natural state of endemic low rainfall, limited water availability and high surface evaporation rates (FAO 2009; Kafle and Bruins 2009). Climate seasonality, particularly hot and dry conditions, short growing seasons in cold zones and high daily temperature fluctuations restrict plant growth. The scarcity of vegetation and obstacles to air movement result in windy conditions causing frequent dust storms with little or no moisture to hold soil particles (Laity 2008).

The great variety of Asian drylands is in part determined by coastal or inland locations, climate patterns, evapotranspiration rates, elevation, latitude and physical characteristics such as mountain ranges. They account for 39 per cent of Asia's land mass, with an area and population greater than African drylands (FAO 2009). This includes arid and semi-arid regions with 600 million inhabitants and dry subhumid zones, home to 800 million people, where increased population, land conversion to new uses (agriculture, urbanisation, resource extraction) and intensified resource use (particularly land and water) are significant issues (Warner et al. 2010). East Asia's Gobi is a cold, mid-latitude, continental desert (Sternberg et al. 2009); Central Asia is a palaeo-arctic, biologically rich arid zone, South Asia is home to the monsoon-influenced alluvial plain of the Thar and neighbouring semi-arid regions whilst the Arabian desert in West Asia features coastal to mountain topography and a hyper-arid interior (Stoppato and Bini 2003; Laity 2008). Each differs in aridity, hazard exposure, resources, infrastructure and tradition yet share similar water, soil and environmental limitations. Social, political and economic factors vary between states, exemplified by government systems (autocracies to democracies), population demographics, development trajectories, conflict and mineral wealth (Haber 2011; CIA 2014). Yet most of Asia's dryland areas are united in limited understanding of hazard dynamics and a history of neglect in the development process.

Globally drylands present a striking paradigm as they are disproportionately home to poor, climate-dependent residents engaged in agriculture and livestock herding (UNCCD 2011). This is stressed in the Millennium Ecosystem Assessment's identification of Asian drylands as the poorest of seven ecological zones, reflected in high infant mortality rates and low GDP per capita (Safriel et al. 2005). Poverty results from several interconnected factors commonly driven by livelihood exposure to environmental forces. Farmers and livestock raisers face climate risk which may disrupt work, reduce income and decrease community viability and resilience. Moreover, expanding populations, intensified land use and changing climate patterns can initiate degradation, reduce ecological productivity and increase vulnerability to hazard events (Warner et al. 2010). These are key concerns; recent research stresses that the social and economic cost of disasters 'are disproportionally borne by poor people in developing countries' (Kellenberg and Mobarak 2008, p. 788). Desertification, a permanent loss of landscape productivity, degrades landscapes and highlights the reinforcing links between poverty and

environmental damage in marginal dryland environments (Reynolds et al. 2007). Vulnerability to physical and socio-political stressors place livelihoods, communities, development and environmental sustainability at the forefront of the climate-human predicament in dryland nations (Scheffran et al. 2011).

Climate hazards

The power of climate hazards to disrupt human and environmental systems is exemplified by drought-induced famines in China that killed 3.5 million people in the 20th century (EmDat 2015). The 1958–1961 episode is particularly infamous; the disaster was initiated by a climatic event and then exacerbated by governance and human action (Bruins and Bu 2006; Dikötter 2010). Floods in 2010 covered 20 per cent of Pakistan and caused €35 billion damage; extreme winter conditions in 2010 killed 25 per cent of Mongolia's livestock and reduced GDP by 4.4 per cent. Omani storms (2007, 2011 – $5 billion damage), China's 2011 drought (>1,291,500km^2 in area) and India's 2002 drought (affecting 300,000,000 people) exemplify the extreme human impact climate hazards can display (EMDAT 2015). These episodes illustrate the potentially far-reaching implications of climate hazards where marginal environments reduce the scope for adaptation and alternative measures (Middleton et al. 2011).

The range and influence of climate hazards is extensive; in drylands slow-onset disasters affect the largest numbers of people and result in the greatest mortality rates (Middleton and Sternberg 2013). Reflected in Asia's major dryland hazards of drought, floods and extreme weather conditions, risk encompasses precipitation, temperature, dust, wind and storm events (Dai 2011). Initiation, severity and spatial and temporal extent can be difficult to identify and quantify for slow-onset hazards in contrast to abrupt episodes (floods, storms). Drought predominates, significantly impacting meteorological (precipitation), agricultural (water, soil moisture) and hydrological (water supply) parameters (Lioubimtseva et al. 2009). Globally, drought is the most damaging hazard with events exacerbated in drylands where endemic low moisture, limited ecological productivity and exposed livelihoods increases vulnerability and makes adapting to negative impacts particularly challenging. Extremes in heat (>50°C) and cold (-45°C) present similar challenges in identification and response.

All climate hazards affect dryland productivity and communities, limit ecosystem services and exacerbate human-induced processes (i.e. resource consumption, environmental degradation, desertification). Changes in climate seasonality, intensity or predictability often result in disproportionate physical impacts (IPCC 2012). For example, shifts in precipitation can have several effects: decreases may instigate drought whilst increases may risk flooding. Fluctuations in precipitation and temperature seasonality are particularly important for agriculture, particularly for planting and vegetation growth. Whilst annual precipitation may appear steady, greater intensity leads to high runoff events and leaves less water to provide soil moisture or groundwater recharge. Drought is not just an agricultural problem; precipitation shortages reduce subsurface water, reservoir levels

and supply for urban centres. These factors can have significant impact on dryland ecology, physical resources and social development. Projected increased aridity, widespread droughts and temperature rises will have serious implications for very large numbers of people in drylands (IPCC 2007, 2012).

The most striking example of hazard-human interaction is the role of drought in the Arab Uprisings and now the collapse of governance in parts of the Middle East (Werrell et al. 2015). A severe 2006–2011 Syrian drought created a 'perfect storm' of physical exposure, socio-economic vulnerability and poor governance that led to high outmigration to cities, then protest, conflict and civil war. Extreme levels of drought in northeast Syria and the hottest temperature extremes in 100 years resulted in 75 per cent crop failure, 85 per cent animal mortality and displaced 1.5 million people (Mohtadi 2012). Loss of livelihoods, inadequate and heavy-handed government policies, selective or lack of response and relief and 2–3 million people experiencing extreme poverty became causal factors of conflict and civil war (Kelley et al. 2015; Werrell et al. 2015).

A second climate example reflects the international scale at which hazards can operate. In 2010–2011, an extreme drought in China's wheat growing region had a major impact on food imports in the water-scarce Middle East (Sternberg 2012). This reduced harvests and threatened supply, leading China to purchase large volumes of wheat on international commodity markets. The process disrupted the wheat fungibility, led to wholesale prices doubling and a reallocation of global supply. The resulting distribution shortages were particularly harsh for importing nations, the top nine all located in the Middle East. Most affected was Egypt, the world's largest wheat importer, where wheat prices tripled and shortages led to bread riots and deaths and merged with political protests. Further, drylands are both high net food importers (Kuwait and Turkmenistan – 96 per cent; Mongolia, Saudi Arabia, Yemen – >80 per cent) and spend much of household income on food (Afghanistan – 61 per cent, Kyrgyzstan – 55 per cent, Pakistan, Syria, Yemen – >42 per cent) (Table 10.2). This process reflects the potential for hazards to have global implications far beyond local or regional scales, the level where hazards are usually examined.

Hazard exposure

Whilst Asia represents rich social, cultural and physical dimensions and diversity, much of the continent is linked by common dry environments, significant physical and social risk levels and large exposed populations. Challenges centre on food, water and livelihoods and are shaped by population, income levels, governance, landscape and climate. The physical dynamics include precipitation, temperature, drought and floods (Table 10.1). Socio-economic parameters reflect human exposure to the environment and disaster risk framed by poverty and exposure (Table 10.2). Most West and Central Asian states are extremely dry (to 99 per cent) regions. Even India and China, often thought of as monsoonal or subtropical nations, in fact encompass vast drylands. Further, whilst there are high temperature clusters in West Asia, the globally coldest states are high latitude Central and East

Table 10.1 Physical data on selected Asian dryland countries (IPCC 2012). Within drylands, the data uses national capitals to site temperature, precipitation and drought data. Temperature reflects average annual and standard deviation (range) from the average. Drought reflects the percentage of time since 1900 in drought, the longest 12-month drought and most recent severe or extreme events. Precipitation identifies mean annual amount since 1900 whilst cities at flood or storm risk are identified. Climate data is from 1900–2013, sourced from the Climate Research Unit, UK, global dataset v3.22 (Harris et al. 2014). Drought is calculated by the Standard Precipitation Index (Sternberg 2012)

Country	*Capital*	*Country - % dryland*	*Average temperature*	*Temperature standard deviation*	*Drought – 12-month scale*			*Capital precipitation avg/year*	*Flood/ storm risk*
					Per cent of time	*Longest in months*	*last severe/ extreme*		
Afghanistan	Kabul	94	10.1	9.0	15	31	2001	495.2	Yes
China	Beijing	52	12.2	10.9	18	29	2010	512.6	Yes
India	New Delhi	69	24.9	6.8	16	23	2002	708.4	Yes
Iran	Tehran	90	12.8	9.7	17	26	2011	300.1	Yes
Iraq	Baghdad	99	22.7	8.9	14	35	2013	140.6	Yes
Israel	Jerusalem	99	18.7	5.6	13	35	2011	473.4	~
Jordan	Amman	99	19.0	5.9	15	37	2011	321.0	~
Kazakhstan	Astana	99	2.5	13.9	15	32	2011	315.0	~
Kuwait	Kuwait City	99	25.6	8.7	11	13	2011	106.1	~
Kyrgyzstan	Bishkek	80	3.9	10.5	14	30	2008	379.5	~
Mongolia	Ulan Bator	80	–2.0	14.2	15	35	2012	277.8	Yes
Oman	Muscat	99	26.2	4.8	15	21	2010	72.7	Yes
Pakistan	Islamabad	80	19.8	7.2	17	29	2001	1075.5	Yes
Saudi Arabia	Riyadh	99	25.4	7.6	15	35	2011	103.0	Yes
Syria	Damascus	98	12.4	6.8	15	27	2011	582.6	~
Turkey	Ankara	82	10.4	8.1	15	30	2007	372.1	Yes
Turkmenistan	Ashkhabad	99	10.4	9.7	13	27	2009	288.4	~
UAE	Abu Dhabi	99	27.2	5.7	14	26	2012	58.4	~
Uzbekistan	Tashkent	99	14.0	9.6	14	55	1996	383.0	~
Yemen	Sanaa	99	15.0	3.2	15	31	1991	464.0	~

Data from Climate Research Unit (Harris et al. 2014), WRI

Table 10.2 Socio-economic factors that frame climate hazard risk in Asian drylands. All countries experience water scarcity. Several exemplify food stress due to amount of cultivated land per person, high rural population, household income spent on food, high food imports and low human development levels

Country	*Country Population (millions)*	*Water scarcity*	*Food – % of income*	*Cultivated land, ha. per capita*	*Food Imports – Net %*	*Rural population*	*Human Development Index*
Afghanistan	32.5	Extreme	61	0.26	70	74	169
China	1,377.0	Medium	36	0.08	30	51	101
India	1,311.0	High	35	0.13	~	69	134
Iran	79.1	Extreme	26	0.24	38	31	88
Iraq	36.4	High	41	0.14	~	33	132
Israel	8.1	Extreme	17	0.04	60	8	17
Jordan	7.6	Extreme	40	0.1	53	18	95
Kazakhstan	17.6	Extreme	37	1.45	32	46	69
Kuwait	3.9	Extreme	15	0	96	2	46
Kyrgyzstan	5.9	Extreme	55	0.24	3	65	126
Mongolia	2.9	High	39	0.16	87	32	110
Oman	4.1	Extreme	30	0.04	55	27	89
Pakistan	188.9	Extreme	46	0.16	30	64	145
Saudi Arabia	31.5	Extreme	23	0.12	81	18	56
Syria	18.5	High	42	0.42	~	44	119
Turkey	78.6	High	25	0.26	~	26	69
Turkmenistan	5.3	Extreme	32	0.35	96	52	102
UAE	9.1	Extreme	9	0.01	95	16	30
Uzbekistan	29.9	Extreme	35	0.21	~	64	115
Yemen	26.8	Extreme	45	0.05	86	68	154

Source: Sternberg (2016)

Asia deserts where Mongolia is home to the coldest capital city. High temperature, climate variability and low moisture limit vegetation and crop production in warm drylands whilst cold environments have low ecological productivity and short frost-free growing seasons. Recent severe and extreme drought in Asia, particularly in 2011 (Table 10.1), emphasise the immediacy of climate hazards throughout drylands. Most nations feature environmentally dependent rural livelihoods and urban centres with high water scarcity and dependence on food imports that place residents at hazard risk (Table 10.2).

Asia's multiple arid and semi-arid regions present different risk profiles with exposure varying with landscape and livelihood. Most vulnerable are rural communities that depend on environment, vegetation, temperature and crop production for income. The largest group are farmers, often with limited arable land (in most countries <0.25 hectare per person), who rely on rainfed agriculture and are exposed to variegated climate factors. The second customary livelihood across

Asia drylands has been pastoralism, herding and settled livestock raising where animals graze natural vegetation and thus rely on pasture and climate for fodder. Climate variability, identified by great yearly and interannual fluctuation in deserts, can be detrimental to conventional livelihoods. Ideas of community adaptability and resilience to hazards depend on the ability to prepare and mitigate disasters, a serious challenge for poor residents. Customary livelihood approaches adapted to climate – mobility, crop selection, traditional water structures – compete with engineered environments, intensified land use, development patterns and policy restrictions.

Whilst the immediate burden of hazards is greater in rural regions, urban centres face indirect climate exposure to water and food supply. Much water is delivered through mega-projects, such as the South to North Water Transfer Scheme in China, vast dam projects such as the GAP in Turkey, canals in India and desalination plants across the Gulf States (Sternberg 2015). These reflect the continual effort to supply adequate water for large urban populations, usually at great financial cost, and are coupled with high groundwater (over)extraction rates. Floods and storms often damage communities and urban infrastructure and, because of intensity, the water is not stored for later use whilst in drought the challenge is adequate water supply. Whereas in rural areas disaster affects cropland and small communities, in cities disaster has high economic and human cost because of population density and development, with recent examples from Pakistan and Oman.

Across the continent hazards are recurring events yet engagement remains an ongoing struggle. The tasks – adequate planning, preparation and response to mitigate hazards and increase resilience – are outlined in the Hyogo Framework. Yet abilities to alleviate hazards depend on economic levels, government attention and capacity, adaptive skills and myriad smaller contributory factors. As highlighted in the Millennium Ecosystem Assessment (Safriel et al. 2005), social vulnerability plays a major role and is framed by poverty and exposure to climate, placing farmers and herders at the forefront of hazard-human interaction. Human dynamics act as stressors – growing populations, competition for land and water, shifts in land use, changing lifestyles and consumption patterns, urbanisation and coping strategies affect hazard impact. Yet there are no desert-wide initiatives or groups for exchanging information or ideas or to establish collaboration on hazards mitigation.

Within countries, dominant perceptions and political realities affect hazard response. China's self-conception stresses the temperate and subtropical central and southern Han heartland within the Great Wall. Framed by high precipitation and warm temperatures, the area is subject to floods and storms affecting a billion people; thus the region receives much government support. Yet the vast arid north and west (52 per cent of land) experience drought and severe winters and problems and challenges more similar to other arid countries than with southeastern China. Thus when productive farmland in the south is converted to urban centres and replaced by cultivation in the north and west, climate exposure transitions and is often increased due to a lack of water sources, poor soils, drought and extreme cold. Also, policies may frame locally adapted strategies as detrimental to national

development. The contrasting hazard scenarios require domestic attention; these are analogous to diverse internal challenges in India's monsoonal south and arid northwest. The significant pan-desert connections and challenges should encourage cross-border discussion and interaction on effective mitigation strategies and measures.

Governance and the state

Government objectives, policy and perspectives are key to hazard management and reduced vulnerability. Control of people, regions and decision-making processes have a major impact on straightforward hazard planning and response. Effective policy can be the most important factor at all development levels in providing direction and support to mitigate hazards; the lack thereof, or failure to engage, can lead to climate disasters. Examples of actions increasing exposure include expanding farmland in dry areas, crop selection, water diversion, reliance on engineered solutions, ignoring environmental conditions, policies to settle 'backwards' pastoralists, location of urban centres and poor disaster planning and response are common factors in most countries. Short-term focus on economic growth that favours political groups and industries, land degradation and lack of investment or interest in some social groups further exacerbates exposure. Asian dryland states have signed the Hyogo Framework yet recent disasters suggest that this has not resulted in widespread improvement in hazard mitigation.

The role of government policy and action, human and financial resources and mitigation capacity differ across arid countries. East, Central, South and West Asian regions differ in aridity, hazard exposure, resources, infrastructure and tradition yet share similar water, soil, environmental and social limitations. Developed nations (the Gulf States, Israel) with advanced technological processes can inform and improve dryland lives beyond national borders. Middle-income countries (China, Iran) may embrace progressive approaches and have capacity whilst also maintaining local methods. Developing states (Mongolia, Jordan) and poor nations (Yemen, Uzbekistan) can benefit from key research, experiences and practices in concert with customary actions. A balance is needed – China's powerful government has shown the ability to respond to disasters, yet environmental degradation, intensified land use and pollution increase vulnerability; Mongolia exhibits natural resilience through mobility but has little response capability. High population density, monsoonal floods and human-induced degradation lead to vulnerability in South Asia whilst population, conflict, water scarcity and reliance on food imports characterise West Asia (Laity 2008). Central Asia states divert the Syr and Amu Darya Rivers for agriculture at a rate that has desiccated the Aral Sea (Sternberg 2015).

A corollary is to examine forms of governance. Deserts feature autocracies – kings and a sultan, one party states, ex-communist dictators, theocracies to democracies and failing states. The variegated ruling systems have a big impact on how policy is set, citizen engagement, priorities chosen, funds spent and programmes implemented. States in the Persian Gulf have used infrastructure and trade to

reduce citizen's climate dependency. Other authoritarian systems (Central Asia) limit open discussion, may be slow to prepare and react and stress the power of nature when climate events strike. Democracies respond to public interests and pressure though disaster response may depend on financial ability and effective implementation. Israel mitigates drought through technology; whilst, after a winter disaster, Mongolia waits for aid, Pakistan seeks international relief to deal with floods and India fitfully copes. Whatever the politics, the key is reducing human risk whether driven by elections, economics or fear of unrest through processes that vary widely across Asia.

As nations pursue different development paths their ability to prepare and mitigate climate events vary greatly. Asian pastoralism presents several approaches to a shared traditional lifestyle that with support can be a source of resilience. In Gulf States, herders are maintained as cultural artefacts, supported by government programmes providing fodder and water delivery, factors that greatly reduce risk. Herding continues in western India, parts of Pakistan, Iran and Iraq and to some degree amongst Bedouin in the Middle East. Often in these regions, hazard vulnerability is framed by poverty, exclusion, lack of government support and marginalisation within the larger society. In Mongolia, still-mobile herders cope with hazards through traditional practices. When winter livestock mortality strikes, it in part reflects limited planning, lack of preparation and the state's inability to mitigate and adequately respond to disasters. The country's recent pattern reflects poor nations' reliance on external assistance for local conditions. Much depends on the government involved – the aforementioned Arab Spring case studies provide opposing examples. Whereas President Assad in Syria did not assist residents affected by drought though a new Drought Management Plan was in place (UN 2009), the Chinese Communist Party made significant efforts to alleviate drought impacts to maintain social stability.

Environment

The IPCC (2007, 2014) documents variability and change across Asian deserts and highlights the exposure of drylands to multiple hazards. In part these are driven by climate dynamics and change, in part by increased human pressure and engagement via population, resource use and need for more productive land. The issue at hand is natural limits in deserts. We are aware of West Asia's water scarcity; more people, more demand for food and changing consumption patterns lead to efforts to maximise land use and exploitation of water sources such as surface water and underground aquifers. This feedback loop then drives the need for water and thus vulnerability to climate hazards, particularly related to agriculture. This is matched by intensified land use that sees more people dependent on marginal landscapes, common across West and South Asia and China.

A second and neglected environmental risk factor is that degradation and over-exploitation of land and water exacerbates vulnerability. Mao famously declared a 'war on nature' that framed China's engagement with the environment whilst the Soviets initiated the 'conquer the virgin lands' programme in an attempt to farm

the Central Asian steppe and desert. Both made land serve the needs of government rather than nature and came to reflect how human action in drylands may reduce productivity and threaten the ability to manage climate disasters. The classic example is the Aral Sea in Central Asia that has been identified as the world's worst environmental disaster. As great amounts of water were diverted for agriculture, natural ecosystem processes were disrupted, changing area weather patterns and increasing exposure to climate throughout the region.

Social impacts

Risk is framed by population, development level, hazard exposure and food security: poorer, rural, agricultural livelihoods and communities are more vulnerable to climate events and have less adaptive capacity. Factors that lead to low Human Development ranking (Table 10.2) – such as income, education, household and institutional capacity and high exposure to climate-dependent livelihoods – limit hazard mitigation. This is amplified in countries with food and water stress, particularly in West Asia, and weak, corrupt or ineffective government institutions, whether democracies or autocracies. The result is vulnerability to hazards with minimal natural ways or societal interventions to reduce risk. The breakdown of governance (Yemen, parts of the Middle East), migration and displacement to cities, increasing poverty and growing food insecurity stress survival and reduce political focus and will to address less tangible hazard risk.

Desert countries are challenged by several common social factors, many of which are exacerbated by dry conditions (Table 10.2). All these countries have high water scarcity, low availability of arable land and a need for vast food imports. When matched with expanding populations and low development levels, climate vulnerability becomes a common stressor. Whilst higher income countries struggle with water and food supply, they have the means to address the issues through infrastructure, desalination, advanced technology, intensive agriculture and have money to import food. Indeed, rich countries may disregard nature – the UAE consumes the most water per capita in the world. It is the uncertainty of water for agriculture and lack of hazard assistance that creates risk for rural populations that predominate in poor countries. These nations also spend a high percentage of income on food and depend heavily on bread for calorie intake. Several states spend >35 per cent of household income on food, in contrast with 9 per cent in the UK.

No one talks to each other

One of the interesting shortcomings of international relations, development and relief is that climate policy, hazard preparedness and disaster aid is delineated nationally rather than based on environments, biomes or common landscapes. This leaves neighbouring nations to cope on their own, limits interaction and support, means states must individually develop capacity and structures and that best practices often stop at borders. As Asian drylands cover the development spectrum,

there is much information, practice and expertise that can be shared. Yet a lack of interaction, knowledge and experience leaves disasters widespread but positive lessons country-specific. The outcome is that transboundary cooperation and skills transfer seldom occurs though the international effect of hazards encourages coordination amongst states.

Oddly, desert nations seldom talk about hazard environments, particularly across regions. Where are international organisations or fora that engage with desert issues? Hazard dynamics are constrained by borders rather than presented as climatological and geographical zones. This means that engagement, mitigation and response occurs on a state by state basis rather than based on spatial scale (drought, temperature extreme), watershed (flood), risk (storm) or linked problems (dust source, area of impact). This isolationist approach results from how nations and borders are conceived, limited capacity, costs, donor organisations' country-specific focus and inertia. Though international efforts, such as the Hyogo Framework, encourage disaster risk reduction, results have not led to effective strategies. Natural partners abound in Asia's trans-border desert regions that share similar threats and episodes as well as social histories, such as West or Central Asia. Across the dryland development spectrum expertise on reducing risk, issue-driven cooperation and skills transfer between arid zones are natural links. For example, factors that reduce hazard vulnerability – finding water, improved irrigation, drought tolerant crops, sustainable land use, rural preparedness – can be transferred to other desert areas.

Discussion/conclusion

The intricate relationship between hazards, environments and society is rich in socio-political implications for citizens and states. Hazards serve as unforeseen trigger events – think of drought and floods affecting water resources, crop production, food supply and instigating migration, civil strife and disrupting governments. As endemic challenges in dry biomes, external climate events impact the resilience of social and physical systems with significant implications for populations and governments at domestic, transboundary, regional and continental scales where exposure threatens livelihoods, communities and resources and economics (Adger 2006; Hochrainer and Mechler 2011). Climate disasters then become part of 'hazard globalisation' as events that occur in one region can have a dramatic follow-on impact across borders and continents. A topical example is Sternberg's (2012) article on 'Chinese drought, bread and the Arab Spring' that stresses links between hazards and multiple systems – threats to food supply and social stability with international trade, human well-being and security.

Engaging with vulnerable systems before disaster occurs strengthens awareness and enables potential mitigation; responding to disaster damage and mortality *ex post facto* – is common yet inadequate in an era of climate change, global warming and more people. In Asia there is limited climate hazard investigation and attention, reflecting the low priority given to hazards relative to other pressing issues and a lack of resources. Currently 95 per cent of disaster funding goes to

post-disaster response with little spent on planning or preparation (Hochrainer and Mechler 2011). Excepting China, climate hazard research and engagement is weak in South and Central Asia whilst there is a dearth of studies in the Middle East and Arabia. The result is that international agencies and donor organisations often set the tone for hazard engagement through control of funding and assistance. Affected countries have limited ability to put forth a local perspective or lack the scientific grounding to argue or advance policy and practice beyond current coping mechanisms. Donor agency focus is on immediate conditions rather than on developing in-depth knowledge or new approaches, exemplified by recent Pakistani floods and Mongolian extreme cold events where relief repeats prior efforts. The knowledge gap creates a negative feedback loop where humanitarian efforts base decisions on existing and past practice rather than on contemporary investigation and quantification of climate hazards.

Improved cooperation across nations and regions can commence with planning and preparedness and address response and system resilience. The constraints imposed (funding, borders, skills) by existing structures (states, international and donor agencies) are poorly suited to today's Asia. Many concepts are holdovers from previous eras when climate was inadequately understood, development and infrastructure was moderate, international organisations were being formed after World War II (UN, aid agencies) and the nation state embodied permanence. Yet, since 1970, Asia's population has doubled (UN 2013); with this, power and economics have transitioned in the region. Whilst the past Soviet vs American and first vs third world global dynamics have faded politically, the rigid configuration of hazards as national rather than geographical events endures. Our greater knowledge of climate and hazard parameters, our growing acknowledgement and appreciation of their disruptive role and our awareness of the significant, direct and collateral damage hazards impose give the impetus for reworking hazard engagement to reflect realities in Asia. Indeed, contemporary hazard crises, as the book title suggests, mark climate as a threat to progress and human well-being in our world.

The vulnerability of deserts to climate hazards can be parsed in many ways. First is to recognise that climate hazard exposure in drylands differs from temperate and tropical zones (Middleton and Sternberg 2013). The core is the reduced adaptive capacity due to marginal environments and physical resource limitations. Societies have been able to build cities and make the desert bloom, but this creates dependence on diverse and often distant water sources and increases exposure to climate. The heightened risk factor means drylands present different vulnerability scenarios than other eco-zones. Further, the theme of marginality runs through deserts in nature and livelihoods as typical dryland dwellers – farmers and livestock raisers – remain dependent on climate, ecology and landscapes. The vast number of dryland inhabitants ensures that climate parameters remain a vital concern.

The immediacy presented by conflict, war, migration and millions of refugees in West Asia are such direct threats that focus on climatic causation is lost. As conflict and turmoil engulf the region less attention is given to the major role of climate and environment in society. Who is growing the food, maintaining infrastructure, developing new water sources in the extensive battle zones? Where are

buffer systems in place to cope with temperature stress or lack of rain, or reduced harvests? How do functioning states like Jordan or Lebanon provide water for 2 million refugees in a sustainable way? War and insurgency present so many challenges that in such humanitarian disasters climate factors are seldom noticed. Yet developing water sources (e.g. Disi Aquifer, Jordan) and the infrastructure to transfer water several hundred kilometres to refugee camps in the desert is expensive to construct. What sounds like a water supply problem is also a climate matter as precipitation ultimately controls water resources, groundwater recharge and the viability of rainfed agriculture. As climate stresses food production whether regions are controlled by governments or the Islamic State, community vulnerability increases commensurately. In the populated, marginal environments moderate changes in precipitation can have exponential impact. In Syria, drought acted as trigger event; what may be the climatic cause of future insecurity?

This chapter started with information about climate extremes, vast Asian deserts, population exposure and rising threats to the environmental and social status quo. Research reflects the variegated dryland conditions, divergent ways to mitigate climate (development, tradition, infrastructure), implied vulnerability of hundreds of millions of poor farmers and animal raisers and the disconnected framework and lack of hazard cooperation and mitigation within and across nations. This unproductive, even dysfunctional socio-climatic landscape ignores both the abrupt power of natural events and the capacity of humans to address and resolve threats to stability and survival. The global risk is in not understanding how changing climates affect societies in marginal environments; the task is to reconfigure climate engagement away from an abstract science to today's imperative knowledge. For states, the chore is to adapt, build resilience and address vulnerability, realising that events can affect political viability, national security as well as regional stability. Syria presents the cost of neglect and emphasises that hazards are delineated by environment rather borders. The personal, community and societal threat is great enough that the imperative is to 'integrate climate change adaptation with disaster risk reduction' (Hyogo Framework/ISDR 2011, p. 10) at national, regional and continental levels. In today's interconnected planet, climate hazards, like people, goods or money, provide new representations of our globalised world.

References

Adger, W.N. (2006). Vulnerability. *Global Environmental Change*, 16(3), 268–281.

Bruins, H.J. and Bu, F. (2006). Food security in China and contingency planning: The significance of grain reserves. *Journal of Contingencies and Crisis Management*, 14(3), 114–124.

CIA World Factbook. (2014). www.cia.gov/library/publications/the-world-factbook/ (Accessed: 1 February 2015).

Dai, A. (2011). Drought under global warming: A review. *WIREs Climate Change*, 2, 45–65.

Dikötter, F. (2010). *Mao's Great Famine: The History of China's Most Devastating Catastrophe, 1958–1962*. Bloomsbury Publishing, London.

EMDAT – International Disaster Database. (2015). www.emdat.be/database (Accessed: 15 September 2014).

FAO. (2009). Drylands, People and Land Use. fao.org/docrep/012/i0372e/i0372e01.pdf (Accessed: 14 September 2012).

Guha-Sapir, D., Vos, F., Below, R., with Ponserre, S. (2012). *Annual Disaster Statistical Review 2011: The Numbers and Trends*. CRED, Brussels.

Haber, Stephen and Menaldo, Victor A. (2011). Rainfall, Human Capital, and Democracy. doi:10.2139/ssrn.1667332. http://dx.doi.org/10.2139/ssrn.1667332

Harris, I., Jones, P. Osborn, T. and Lister, D. (2014). Updated high-resolution grids of monthly climatic observations–the CRU TS3. 10 dataset. *International Journal of Climatology*, 34(3), 623–642.

Hochrainer, S. and Mechler, R. (2011). Natural disaster risk in Asian megacities: A case for risk pooling? *Cities*, 28(1), 53–61.

International Strategy for Disaster Reduction (ISDR). (2009). *Global Assessment Report on Disaster Risk Reduction*. United Nations, Geneva.

International Strategy for Disaster Reduction (ISDR). (2011). Hyogo Framework for Action 2005–2015: Building the Resilience of Nations and Communities to Disasters. www.prevention web.net/files/18197_midterm.pdf (Accessed: 25 January 2012).

IPCC (Intervernmental Panel on Climate Change). (2007). *Climate Change*. Cambridge Press, Cambridge.

IPCC. (2012). *Field et al.* Cambridge Press, Cambridge.

IPCC. (2014). Climate Change 2014: Impacts, Adaptation, and Vulnerability. 23 March 2014. www.ipcc-wg2.gov/AR5/report (Accessed: 25 January 2015).

Kafle, H.K. and Bruins, H.J. (2009). Climatic trends in Israel 1970–2002: Warmer and increasing aridity inland. *Climatic Change*, 96(1–2), 63–77.

Kellenberg, D.K. and Mobarak, A.M. (2008). Does rising income increase or decrease damage risk from natural disasters? *Journal of Urban Economics*, 63(3), 788–802.

Kelley, C.P., Mohtadi, S., Cane, M.A., Seager, R. and Kushnir, Y. (2015). Climate change in the Fertile Crescent and implications of the recent Syrian drought. *Proceedings of the National Academy of Sciences*, 112(11), 3241–3246.

Laity, J. (2008). *Deserts and Desert Environments*. John Wiley & Sons, Chichester.

Lioubimtseva, E. and Henebry, G.M. (2009). Climate and environmental change in arid Central Asia: Impacts, vulnerability, and adaptations. *Journal of Arid Environments*, 73(11), 963–977.

Meigs, P. (1953). World distribution of arid and semi-arid homoclimates. In *UNESCO Reviews of Research on Arid Zone Hydrology*. United Nations, Paris. unesdoc.unesco.org/images/0014/001485/148501eb.pdf

Middleton, N., Stringer, L., Goudie, A., and Thomas, D. (2011). The forgotten billion: MDG achievement in drylands. In: *United Nations Convention to Combat Desertification, Bonn.* UNDP-UNCCD, NY.

Middleton, N. and Sternberg, T. (2013). Climate hazards in deserts: A review. *Earth Science Reviews*, 126, 48–57.

Mohtadi, S. (2012). Climate change and the Syrian uprising. *Bulletin of the Atomic Scientists*. http://thebulletin.org/climate-change-and-syrian-uprising (Accessed: 15 September 2014).

Reynolds, J.F., Smith, D.M.S., Lambin, E.F., Turner, B.L., Mortimore, M.M., Batterbury, S.P.J., . . . Walker, B. (2007). Global desertification: Building a science for dryland development. *Science*, 316, 847–851.

Safriel, U., Adeel, Z., Niemeijer, D., Puigdefabregas, J., White, R., Lal, R., . . . King, C. (2005). Chapter 22: Dryland systems. In Hassan, R., Scholes, R., and Ash, N. (Eds.) *Millennium Ecosystem Assessment: Ecosystems and Human Well-Being*. World Resources Institute, Washington, DC. 623–662.

Scheffran, J. and Battaglini, A. (2011). Climate and conflicts: The security risks of global warming. *Regional Environmental Change*, 11, 27–39.

Sternberg, T. (2012). Chinese drought, bread and the Arab Spring. *Applied Geography*, 34, 519–524.

Sternberg, T. (2015). Water megaprojects in deserts and drylands. *International Journal of Water Resources Development*, 32(2), 1–20.

Sternberg, T. and Batbuyan, B. (2013). Integrating the Hyogo Framework into Mongolia's disaster (DRR) policy. *International Journal of Disaster Risk Reduction*, 5, 1–9.

Sternberg, T., Rueff, H. and Middleton, N. (2015). Contraction of the Gobi desert, 2000–2012. *Remote Sensing*, 7, 1346–1358.

Sternberg, T., Thomas, D. and Middleton, N. (2009). Pressurized pastoralism in South Gobi Province, Mongolia: What is the role of drought? *Transactions of British Geographers – IBG*, 34, 364–377.

Stoppato, M. and Bini, A. (2003). *Deserts*. Firefly Books, Toronto.

UNCCD. (2011). Global Drylands: A UN System-Wide Response. www.unccd.int/Lists/Site DocumentLibrary/Publications/Global_Drylands_Full_Report.pdf.

United Nations (UN). (2009). Syria Drought Response Plan. http://docs.unocha.org/sites/dms/CAP/2009_Syria_Drought_Response_Plan.pdf (Accessed: 22 January 2012).

United Nations. (2013). Population division. *Department of Economic and Social Affairs*. www.un.org/en/development/desa/population/ (Accessed: 25 January 2014).

USDA (United States Department of Agriculture). (2011). www.usda.gov/wps/portal/usda/usdahome; www.indexmundi.com/agriculture/?Country¼eg&;¼wheat&graph¼ imports (Accessed: 30 April 2012).

Warner, K., Hamza, M., Oliver-Smith, A., Renaud, F., and Julca, A. (2010). Climate change, environmental degradation, migration. *Natural Hazards*, 55, 689–715.

Werrell, C.E., Femia, F. and Sternberg, T. (2015). Did we see it coming? State fragility, climate vulnerability, and the uprisings in Syria and Egypt. *SAIS Review of International Affairs*, 35(1), 29–46.

White, R. and Nackoney, J. (2003). *Drylands, People, and Ecosystem Goods and Services*. WRI, Washington, DC.

World Bank. (2014). Urban Population. http://data.worldbank.org/indicator/SP.URB.TOTL.IN.ZS (Accessed: 24 February 2015).

11 Climate change

Rethinking the local for policy and practice

Lena Dominelli

Climate change is affecting all peoples on the planet, albeit in different ways. Responding to the challenges of those affected by the 'natural' disasters that follow in its wake is a challenge for practitioners working with victim-survivors on the ground and policymakers who need to undertake advance planning. To begin with, a 'natural' disaster has a (hu)man-made dimension that challenges traditional divisions about disasters made in emergency planning. Interdependencies between countries and people add another layer of complexity in responses to adversity. For example, in the Asia-Pacific region, some small Pacific Island Nations that have contributed little to greenhouse gases are sinking into the Pacific Ocean as industrialised and emerging economies increase their pollution levels and contribute to melting the Arctic and Antarctic ice-sheets, thereby causing ocean levels to rise. Such events ought to shift power relations between the global and the local, especially at the levels of moral and political power. That it has not is extremely worrying, especially in the context of climate change which is expected to increase the intensity and frequency of disasters linked to extreme weather events including those of drought, flooding, tropical storms, heatwaves, cold snaps and wildfires (IPCC 2014).

Challenges to this shift need or have to be embedded in egalitarian and equitable directions if those larger nations with muscle are not to overwhelm countries that are relatively less powerful and currently carry less weight in global decision-making settings. This is an issue of significance to Small Island Nations that have to influence larger ones in international fora including the United Nations (UN). Additionally, the local provides the point at which practitioners tackle the impact of disasters upon peoples and communities. Given the diversity and divisions that often exist within communities, this is not always easy. What can be learnt from previous responses to disasters and experiences to assist practitioners and policymakers in this regard?

In this chapter, I examine reactions to the 2004 Indian Ocean Tsunami in Sri Lanka and the 2013 Typhoon Haiyan that devastated the Philippines. I consider how the local can be reconceptualised to address the interdependencies between peoples and countries at local and international levels more effectively. In the process, I will consider not only practice interventions linked largely to the provision of humanitarian aid, but also policy initiatives concerned with risk, adaptation,

prevention and reconstruction that can be pursued to bring about a more egalitarian re-conceptualisation of the issues that must be dealt with if the world, its peoples, flora, fauna and physical environments are to be preserved for future generations as well as cater for current needs. To achieve these goals, I will draw upon existing literature and also several research projects[1] that I have undertaken, one dealing with the 2004 Indian Ocean Tsunami (Dominelli 2013a, 2014), one involving supporting victim-survivors of typhoon Haiyan (Yolanda) in the Philippines and another associated with the impact of climate change on services for older people in the UK (Oven et al. 2012; Dominelli 2013b).

Humanitarian aid: a controversial issue interlinking the local and the global

Climate change is credited with increasing the frequency and intensity of disasters, primarily those associated with extreme weather events such as extreme heat or cold (IPCC 2014). The UNISDR Disaster Report of 2013 has estimated damages caused by disasters globally at US$1.7 trillion, deaths at 1.2 million and people affected at 2.9 billion. This represents an enormous cost to people and the planet. The Asia-Pacific region as one of the most severely affected by so-called natural disasters ranging from flooding, landslides, tsunamis, earthquakes and volcanic eruptions, is particularly vulnerable, a vulnerability exacerbated by climate change (McGuire 2014). Moreover, it houses over half of the world's population in countries that include the wealthy 'Western' states of Australia and New Zealand, the developed economies of Japan and Singapore, the emerging economies of China and India, and some of the poorest countries in the world, e.g. Nepal, Bhutan and the small island states in the Pacific, each with its own catalogue of risks and susceptibilities.

In 2011, Sanjay Srivastava, the UN's Regional Adviser for Disaster Risk Reduction in Asia, claimed that 90 per cent of the economic losses due to disasters in that year occurred in that region. Moreover, increases in earthquake activity and related hazards, and volcanic eruptions are being attributed to increases in greenhouse emissions and global warming (McGuire 2014), and will cause further havoc as anthropomorphic activities increase (Crutzen and Schwägeri, 2011). As a result, the implications of activities occurring in one location but holding consequences for another have to be addressed. A clear and dramatic illustration of this interlinkage was the 2010 Eyjafjallajökull volcanic eruption in Iceland which grounded air traffic into and out of Europe and the rest of the world from April 10–15, 2010.

Airplane companies, having learnt about the damage fine volcanic ash caused to jet engines from previous volcanic eruptions, including that of Pinatubo in the Philippines in 1991 (Casadevall 1993), responded with extreme caution and refused to fly through European airspace. Eyjafjallajökull also exposed the interconnected nature of economic and natural events, an issue long ignored as part of modernity with its silo-based mentality in which professionals from one academic discipline or sphere of practice rarely conversed with each other. Moreover, as activity in the Eyjafjallajökull volcano is believed connected to a potential eruption in a nearby

volcano called Katla (Larsen 1999), the Icelandic government has already warned airline authorities globally to prepare plans for its possible eruption in the near future because it would carry greater catastrophic potential than the Eyjafjallajökull in 2010. However, the timings of actual eruptions are impossible to predict, and this makes taking targeted preventative action extremely difficult, especially for politicians concerned with short-term electoral cycles (Possekel 1999). Such events call for a rethinking of the relationship between the local and the global because what happens in one country or locale can have serious repercussions for another some distance away.

Until recently, the mainstream globalisation literature treated the local as dependent upon the global, with economic macro-level developments driving what happened at the meso- and micro-levels, especially in communities. The term globalisation is controversial and lacks an agreed meaning (Hirst et al. 2009). In contrast to the one-way street implied in hegemonic definitions of globalisation, I have tried to retheorise the inter-related nature of the global and the local as a two-way dialectical exchange process by referring to their interaction upon and with each other in all aspects of everyday life practices (Dominelli 2004, 2009). The mainstream approach of treating the local as reliant upon the global has resulted in an under-valuing of the impact of the local upon the global.

Humanitarian aid provides another arena in which the local and the global interact, usually in complex ways because global players or responders come to the aid of local victim-survivors of a disaster, whether natural or (hu)man-made. Humanitarian aid is generally defined as assistance provided during a crisis that requires external resources and interventions for a particular affected community to cope. This treats disasters as an aberration from the norm, an outdated view that has been increasingly challenged by complexity theorists (Hilhorst 2003). Such interventions aim to save lives, reduce risk of further damage and enable people to recover quickly or become more resilient than they were before the disaster (Manyena 2006). Aid donors can also take advantage of globalisation's space-time compression (Harvey 1989) to reach affected places quickly. Humanitarian aid involves national governments, multilateral agencies often underpinned by government funding, e.g. UNHCR (now known as the UN Refugee Agency), and nationally and internationally-based NGOs like the International Federation of the Red Cross and Red Crescent Societies (IFRC) who seek donations from governments and individuals. Again, such interventions tend to treat the local as dependent upon the international, despite extensive criticism of this state of affairs (Hancock 1996).

Money is usually given for humanitarian purposes immediately following a disaster – natural or (hu)man-made, often at the expense of longer-term reconstruction and development efforts (Bookstein 2003). Specific vulnerabilities in the Asian region exacerbate its disaster-proneness and include:

- rapid population growth which is increasing at above world average rate in some countries, has rising numbers of older people aged 65 and over and an imbalance between the genders, and has placed pressure on land, food

and energy resources according to the United Nations Department of Economic and Social Affairs, Population Division.

- continued mass urbanisation including the growth of slums which place people in unplanned and unsafe locations as is currently occurring in Dharavi, Mumbai (Cooper 2014).
- the spread of hyperurbanisation, especially in China (Dominelli 2012).
- climate change and its effects on sea levels, global rainfall and storm patterns – climate-related disasters could affect 375 million people every year by 2015, up from 263 million in 2010, according to British MP, Justine Greening.
- chronic hunger affecting millions which is greatly exacerbated by floods and droughts which further reduce agricultural capacity, particularly in Pakistan and Bangladesh, two countries assisted by the UK's Department of International Development (DfID) to recover through development aid in the form of seeds.

Governments and humanitarian aid agencies are now beginning to look for ways of encouraging development that enables people to look after themselves rather than rely on aid handouts (Action Aid n.d.).

The humanitarian field is also a crowded one and poorly coordinated, with duplication of efforts and competition for funding that exacerbates instability and uncertainty in planning interventions (Bennett and Daniel 2002). Other authors are concerned about the politicised nature of humanitarian aid and argue that donor conditions and behaviours skew the humanitarian effort in deleterious directions (Hancock 1996), while some writers contend that humanitarian aid workers are compromised by becoming embedded within military circles and political conflict (Hoogvelt 2007), despite the Red Cross' age-old injunction of retaining neutrality and independence in complex political situations. Pilger (2005), commenting upon the 2004 Indian Ocean Tsunami, complained about the perfidious nature of government promises to deliver aid that never arrived.

Humanitarian aid is delivered by practitioners, volunteers and formal responders at all stages of the disaster cycle from prevention, immediate relief and recovery, and long-term reconstruction, although different agencies tend to focus on one stage more than another. Usually, the immediate relief and recovery phase is the one that engages more external NGOs than the other stages. Relief coordination is important throughout the disaster cycle and includes protection and support services, and communication networks. Immediate relief and recovery usually involves the provision of shelter, water, food, medicines and family reunification services. Most aid given for immediate relief ends within six months. The online survey in the *Internationalising Institutional and Professional Practices* (IIPP) Project focussing on the 2004 Indian Ocean Tsunami (1) revealed that most agencies planned to leave the disaster scene within the relief and recovery period (Medford and Dominelli 2013).

Reconstruction and rehabilitation includes the repair of existing built infrastructures, and in many cases, depending on funding available, their improvement for

greater resilience in future. This stage may last years. Atypically, however, following the Wenchuan Earthquake of 2008, the Chinese government replaced most of the destroyed road, power and communications infrastructures and buildings during the first year after the event. In many circumstances, including the 2004 Indian Ocean Tsunami, multinational firms often lead reconstruction efforts as well as other aspects of the disaster intervention cycle (Klein 2008). Their activities further integrate the local economy into the global one, but in a relationship in which the global usually dominates developments in the local sphere. Moreover, such interactions produce vulnerabilities which underpin understandings and calculations of risk (Blaikie et al. 1994; Pelling and Uitto 2001) and give rise to further complexities on the ground.

Disaster preparedness matters both before and after a disaster strikes and is crucial to mitigating risk, reducing damage when a calamitous event occurs, and enabling people to look after themselves as much as possible during a disaster, particularly before external help arrives. This has been a lesson learnt in Asia following the 2004 Indian Ocean Tsunami, and also Hurricane Katrina in the United States in 2005. External assistance may take some time to reach victim-survivors, so that local people – family and neighbours – usually become the first responders, as many Asian victim-survivors have discovered in earthquakes, volcanic eruptions, floods, typhoons and other disasters. This insight is particularly important in countries with poor transportation and communication infrastructures, e.g. Pakistan, Nepal and the Philippines in Asia. In the West, the American FEMA (Federal Emergency Management Agency) advises people to plan on being self-sufficient for at least 72 hours following a calamitous event. Yet, unlike some others, the United States has an extensive formal helping network to draw upon. Preparedness initiatives include education and training to develop disaster awareness, early warning systems, stockpiling food and supplies, including the compilation of an emergency kit, and depending on the type of hazard, building earthquake-proof and flood-proof buildings and disaster-proof infrastructures especially in power, communications, and transportation to reduce risk and minimise damage. Practitioners often focus on briefing people on early warning systems and building more solid disaster-proof housing to help themselves withstand some of the risk associated with 'natural' disasters.

These elements combine to produce extensive complexities, especially in the relationship between nature, the environment and society. However, Hilhorst (2003) cautions against thinking of this relationship as one operating between systems and sub-systems on the grounds that this would presuppose causality and predictability between them, when this is far from being the case. Human agency, unpredictability in human behaviour, diverse interpretations of events and uncertainty prevent the realisation of the certainty that a systems model assumes. Such models also ignore the human capacity simultaneously to belong to multiple systems and sub-systems and act in fluid and dynamic ways. Thus, modellers can over-estimate the commonalities or universalising aspects of and under-play diversities within systems. Also, systems have dimensions that interact within them as well as between other systems, to add further layers of complexity.

To overcome the problems associated with neglecting complexities, Hilhorst (2003, p. 41) posits reconceptualising the issue in terms of social domains which she defines as 'areas of social life that are organized by reference to a series of interlocking practices and values.' These provide spaces where notions of risk are exchanged as practices, including discursive ones that are not functionally related. This conceptualisation allows for contradictions, conflicts and negotiations to emerge through the interplay of the different elements involved, albeit the utilisation of the same expressions or language can disguise considerable differences, particularly in the interpretations and meanings individuals give to specific events. Hilhorst's conceptualisation allows for social change to occur from within communities as well as being externally imposed. It also facilitates the blurring of boundaries between ideas, people and resources so that the different domains can accommodate complexity in adverse situations.

Hilhorst (2003) postulates that there are three main social domains that feature in responses to risk and disaster. These are the domains of:

- expert-driven science and disaster management;
- disaster-governance driven by bureaucrats and their political controllers; and
- local responses produced by vulnerable people locally, each of which depicts specific ideas about nature-society interactions, vulnerability, risk, disaster response, discourses, meanings and levels of complexity associated with these.

More than 30 disciplines engage in academic discourses around disasters (Alexander 1997) which now include social work (Dominelli 2012), thus adding to the diversity of voices both within and between disciplines. Asia, with its vast expanse of disaster-prone areas and active NGO community involved in disaster mitigation has been chosen as a site for innovative interdisciplinary research in disaster risk reduction that straddles the physical and social sciences. The *Earthquake without Frontiers* (EwF) Project that includes India, China, Nepal, India and Kazakhstan as key sites to examine seismological hazards and the secondary hazards such as landslides and floods associated with these, is one such project. These projects, each partnered by university academics, practitioners and policymakers with local and regional NGOs and local residents to coproduce new knowledge in a new approach to science for use in disasters (Lane et al. 2011).

Hilhorst (2003) identifies the *expert, scientific-management paradigm* as having a binary view of nature, i.e. natural resources that are ripe for exploitation and control through the administration of expertise. This is a hazard-centred approach whereby a hazard must be physically controlled to prevent nature from expressing its inherent power. It is typical of modernist attempts to channel nature and resources into particular directions to maximise profits and occurs through the processes of commodifying physical and environmental resources. This understanding of disasters is dominated by geo-physical approaches that monitor events and seek to improve predictability by engineering solutions that will control nature, e.g. flood embankments. The weaknesses of this model are exposed when nature transcends the limitations of engineered solutions, a feature that is beginning to

raise questions about this way of solving the problems encountered in nature-society interactions (Lane et al. 2011). An example of this in Asia is China's Three Gorges Dam mega-project which has given rise to 'dead' water in the river and droughts in once fertile agricultural lands. In its construction, local residents' arguments for small, linked reservoirs that would better meet their needs were disregarded (Aird n.d.).

The expert scientific-management model has been extended through the *disaster-governance one* that focuses on administering and managing disasters through disaster reduction plans, and pre-prepared emergency responses that are often left in the hands of the military to implement as the first-respondents in a calamity (Hilhorst 2003). This paradigm assumes that interventions occur largely as linear processes that divide complex issues into independently manageable parts. More recently, under pressure from actors based in the margins of society, the disaster-governance model has begun to consider sustainable solutions to disasters rather than engineering feats, especially in those related to flooding. For example, dike construction in the Netherlands has given way to water meadows, in a solution traditionally more palatable in the Global South than in the Global North (Rijkswaterstaat 2002).

Disaster-governance paradigms are used to define vulnerability and risk through techno-bureaucratic processes and understandings that are mediated through specific social institutions to provide particular interventions that are regulated through state action. These are further impacted upon through local or indigenous narratives of governance and risk that are culturally bound because certain features are selected as being more relevant than others to the technocratic narrative being expounded by bureaucrats (Douglas and Wildavsky 1982). However, these authors' view that there is a singular culture of risk has been undermined by the diversity of discourses, explanations and responses around disasters and risk that simultaneously contend for public space and acceptance (Hilhorst 2003). An interesting issue in this regard in Asia is the role of historical culture, traditions and values, which are often argued as being different from the West's (Yip 2005).

Internationalisation complicates the disaster-governance paradigm further by adding international levels of governance and regulation. While the bodies determining these rules and regulations may be dominated by more powerful countries, e.g. the Security Council in the United Nations, the idea of taking action at these levels is appropriate given that disasters do not recognise the national boundaries created by specific nation-states. However, many international interventions are externally led and lack locality specific knowledge. Its absence may add further instability and lack of coherence to intervention strategies and consequently worsen poor forms of coordination already in place. Marginalisation can intensify when local bureaucrats add further conditions of eligibility that can further exclude marginalised groups, thereby exacerbating existing vulnerabilities. '[P]atterns of risk governance evolve in the everyday practices of risk and disaster management' and aggravate state-society-individual relations around expectations that the state would protect its citizens from all forms of potential harm (Hilhorst

2003, p. 45). The mismatch between expectations and their realisation can result in social change that can: 1) facilitate democratisation, 2) reinforce prevailing patterns of inequalities or 3) buttress existing military dictatorships. The direction that change takes is shaped by the interactions that occur between scientists, managers and local, often marginalised, residents in the institutional spaces of the disaster-governance paradigm.

The *locally produced disaster-response* paradigm draws upon people's local or indigenous knowledge, skills and capabilities (Hilhorst 2003). Self-help among neighbours and kin becomes the major basis of support. Duffield (1993, p. 144) pointedly highlighted that only 10 per cent of survivors in emergencies can attribute their outcome to external sources of relief, thereby highlighting the importance of local neighbourhood and kin-based responses, particularly in Asia where publicly funded forms of intervention are rare. Knowledge can only be considered 'local' by external agents who consider their own knowledge as universal, and therefore, 'superior.' Jansen (1998) divides local knowledge into 1) utilitarian or instrumental approaches wherein local understandings can be drawn upon for disaster management purposes, 2) holistic approaches that eschew the binary division between nature and culture in an attempt to separate locally determined action from that driven by the ideas of modernity and colonisation which emanate overseas (Harvey 1990) and 3) self-empowering approaches that draw on local knowledge and capacities as sources of political-economic growth that is based on an alternative (to mainstream) development agenda by encompassing self-reliance, ecological soundness and popular empowerment. Hilhorst (2003, p. 48) considers this latter view one of the 'structuralist approaches to disasters calling for participatory societal change to structurally address vulnerability,' an approach that is gaining favour among NGOs like Action Aid, Oxfam and Save the Children.

This local approach allows for differentiation among community members in knowledges as well as other capabilities and the existence of diverse and contradictory views among them. Arce and Long (2000, p. 24) argue that knowledge should not be conceptualised as a 'binary opposition between Western and non-Western epistemologies and practice . . . [but as] the intricate interplay and joint appropriation and transformation of different bodies of knowledge.' Thus, changes in them occur simultaneously as a result of their interaction and enables local residents to experiment and improvise as necessary in adapting to their conditions, using whatever resources are available to them in utilising their own social capital, social networks, entrepreneurship and externally generated information and technologies. In dynamic, heterogeneous communities, local knowledge 'is not a stock of knowledge but constantly evolves through the social negotiations, accommodations, exchanges and power struggles of local actors' (Hilhorst 2003, p. 49). Moreover, their 'local' inventions acquire authority when supported or accepted by others. In this sense, local knowledge does not differ from scientific knowledge which also relies on peer approval for legitimacy. And, it shifts power away from the experts.

To Hilhorst's (2003) social constructionist approaches in the local production paradigm, I would add what I call the socio-biosphere arena of socio-natural

disasters, or the co-construction of knowledge and solutions to disasters. I advocate this stance to ensure that there is a holistic understanding of the complexities of disasters that combines both expert (often external) knowledges and skills with local and/or indigenous knowledges and skills to ascertain what is locality specific and culturally relevant in particular circumstances. This approach gives human agency a key role in both creating knowledge and power being (re)created in and through that knowledge to reach mutually acceptable solutions.

The Indian Ocean earthquake and tsunami of 2004

The 2004 Indian Ocean Earthquake and Tsunami of 2004 devastated 12 countries bordering that body of water and killed nearly 300,000 people on December 26. Indonesia, Thailand, Sri Lanka and Tamil Nadu in India bore much of this devastation caused by an undersea megathrust earthquake of magnitude 9.1–9.3 (Lay et al. 2005). The tsunami that accompanied this earthquake had waves that rose to 30 metres in height, which destroyed buildings with the ease of breaking matchsticks. The extent of this disaster was overwhelming and it remains one of the disasters with the highest number of casualties of modern times. Its unfolding generated an immense and unprecedented wave of global goodwill that drew in thousands of donors and volunteers, many of whom had never engaged in humanitarian aid before. These people went to help clear debris, provide immediate emergency supplies of water, food, clothing and temporary shelter. People's generosity outstripped that of their governments who had to play 'catch up' by pledging to match private and individual contributions (Pilger 2005). Many official pledges remain to be fulfilled as governments began to backtrack on their promises once the media spotlight moved on. And even today, thousands of people remain in temporary housing despite promises to the contrary. Findings from the IIPP Project revealed that significant numbers of victim-survivors did not receive their share of aid while others were unduly given assistance. This indicates a failure of donors to ensure that those entitled to aid actually received it (Dominelli 2014). Osterhammel (2005) claims that external colonialism utilised globalising assimilationist policies, processes and practices that exploited indigenous peoples, expropriated global resources and promoted ideas of Western superiority. These practices are currently evident in many humanitarian situations (Hancock 1996; Hoogvelt 2007).

The *Internationalising Institutional and Professional Practices Project* (IIPP) funded by the Economic and Social Research Council in the UK focused on the 2004 Indian Ocean Tsunami in 12 villages in Sri Lanka and explored different models of aid to identify the criteria that local people used to evaluate donor-provided services and determine which organisations related to them in empowering ways. This study drew upon an ethnographic methodology to ascertain how players in different locations attached meaning to the disaster, responded to the challenges it presented and influenced each other. Ethnography can reach beyond the local arena to become equally insightful and suitable in and between the various levels and actors involved in disaster responses, exposing both similarities and differences between them. The IIPP also focused specifically on two models of

interventions, the IM (Institutional Model, which began in a British university) and the PPM (Professional Practice Model, which was commenced by an international NGO holding a remit for social work and was followed through by an Eastern European university).

These two models were chosen because they advocated the use of local organisations and local people's knowledge, skills and capacities in developing responses to the tsunami in all phases of the disaster (Dominelli 2014). Their key criteria of operation as identified and considered acceptable by the villagers that were interviewed were as follows. Their emphasis was on self-empowerment through relationships with other organisations whether these originated within the local scene or overseas. Moreover, the villagers' comments endorsed Hilhorst's (2003) local production paradigm. Their key points were:

1. Having links with local people, organisations and communities to ensure that aid delivery processes were and remained locally driven;
2. Involving local people in making decisions about what would happen;
3. Favouring reciprocated, grassroots-based mutual exchanges over top-down, donor driven ones;
4. Having explicit criteria for receiving aid, making these known, applying them fairly and monitoring their outcomes;
5. Ensuring all those requiring aid received it, but not giving aid to the unentitled;
6. Having transparent and efficient aid allocation and distribution systems to avoid nepotism;
7. Encouraging training and raising awareness about disasters including the risks posed and preventative measures that could be taken;
8. Conducting practice drills;
9. Promoting empowering processes and power-sharing between overseas and local organisations throughout the aid distribution process and the reconstruction cycles; and
10. Treating humanitarian giving as a constantly evolving, adaptable and resilient process.

(Dominelli 2014)

Most villagers were appreciative of the aid received, especially in the immediate aftermath of the disaster when the focus was on the provision of food, water, medical supplies, clothing, shelter and reunification services. Below is a typical observation made by one of them:

> Children suffered a lot, they not only lost their belongings, and some of them lost one or both parents. There were children who were caught by the waters, and were injured. Some of them could not sleep at night. They were scared to leave their mothers. Some of the people who came, especially from Red Cross, World Vision, FORUT and Christian Children's Fund did some programmes for the children. They made them play games. Made them to draw, sing and dance. This was very good. Some organisations came and took the

> children to the beach and they played with the children on the beach. This helped the children. Many organisations and individuals gave the children clothes, school equipment, milk foods and other things useful for the children. Some school children even gave the children notebooks with their notes.
>
> (Villager interviewee, IIPP project)

However, the villagers criticised both models for not having sufficient resources to develop sustainable long-term initiatives. The IM was also critiqued for not considering what they could do to ensure that the preschool children had facilities to attend once they outgrew the nursery schools that it had provided. Both the IM and PPM are addressing these issues by undertaking specific capacity-building initiatives driven by local people and organisations.

Although they excluded the IM and PPM from this pronouncement, many villagers were also critical of both overseas and in-country donors who delivered inappropriate aid – duplicating materials given; providing articles that were neither wanted nor necessary and having bureaucratically driven processes, a finding echoing the features dominant in Hilhorst's (2003) scientific-management paradigm. One specific concern that villagers expressed about the culturally inappropriate overseas aid provided was that Sri Lankans had begun to lose their independence and capacity for self-sufficiency by becoming dependent on external providers. One villager articulated this point as follows: 'during the post tsunami period, most of the people had to be helped to get away from the begging bowl attitude.' An interviewee responsible for training social work students puts the emphasis on self-determination slightly differently:

> They [villagers] think, Oh she is from this place. It's something to do with social welfare, so they will donate us something. That's the culture of Sri Lanka. That's the culture because there's lots of foreign aid and things. So, people think when you place [social work] students they're going to give something to us. So what we teach our students is to explain to them first what social work is. So we tell them to tell the villagers, Okay, I'm not here to provide you with anything. But, we, as a village, will get together, talk about your needs and then we will work towards a plan to achieve what you want, to get what you want. So that's normally the concept we try to build in villages.
>
> (Social work trainer interviewed, IIPP project)

Another strong criticism of local, officially driven in-country aid was that the police and military often reproduced existing inequalities by giving assistance to those they knew or who had important connections to local and national dignitaries, regardless of whether they had actually been affected by the tsunami. This empowered bureau-technocratic forms of governance that left victim-survivors feeling disenfranchised and powerless. As one villager graphically explained:

> People trying to get help by producing false [fabricated] evidence. Some people produced false certificates showing that they were tsunami victims even

> though they had not been affected by the tsunami. They 'bribed' officers in state services, e.g., the Grama Niladaris, or other government officials.
>
> (villager interviewed, IIPP Project)

Fortunately, having identified such problems, the government was taking steps to ensure that a regulated and technocratically administered approach would prevail. The argument became that such occurrences would be less likely to happen in future because Sri Lankans will be better informed and more prepared for disasters. And their bureaucratic structures would be more robust and appropriately governed. As one official explained:

> Now we have a very well-equipped [National] Disaster Management Centre. We have a very well-trained entity [and people] who are capable of providing essential services. They are very knowledgeable about their communities. They are fully prepared to respond to any disaster. WE [interviewee emphasis] are ready with all the statistics of the villages under threat, the number of people in each village, their background. All the data is available with us. People are trained and they know how to respond to any disaster.
>
> (District official interviewed, IIPP Project)

This particular statement aims to inculcate credibility and legitimacy in future, government-led responses to calamitous events and endorses Hilhorst's (2003) disaster-governance paradigm.

Additionally, the gender bias against women, first identified by Enarson and Morrow (1998) with regards to Hurricane Andrew in the United States, remains a grave concern and was quite prominent in the 2004 Indian Ocean Tsunami in many organisations (Pittaway et al. 2007; Dominelli 2012). Women were at greater risk through all stages of the disaster, and were often excluded from receiving aid or making decisions about disaster risk reduction and interventions. Ironically, despite the exclusion of women on a gendered basis, poor men also received short shrift through the aid processes. Their specific needs as men were often ignored (Dominelli 2014). In the IIPP project, one villager commented with regards to this:

> I don't think it [aid] has helped the men to change. But for women and children, it was good as they got the opportunity to send their children to a preschool. Women had the opportunity to save, get loans and start self-employment. This is a good thing.
>
> (Villager interviewed, IIPP Project)

The women, in contrast, developed the confidence to arrange a visit to the Temple of the Tooth (a Buddhist shrine) in Kandy through their own initiatives, having come together as a group in an income generation project, initiated by empowering donors.

The 2004 Indian Ocean Tsunami has provided villagers with the skills to take control of situations more effectively in current disasters, especially flooding and

droughts, e.g. by working with overseas practitioners and academics to grow drought and flood-resistant crops, something that is likely to become more important under conditions of climate change. The IM included overseas and Sri Lankan agricultural experts working together as partners. The dominance of the global over the local is also evident in climate change talks held under the auspices of the UN Framework Convention on Climate Change (UNFCCC) where the long-industrialised developed countries identified under Annex 1 of the Kyoto Protocol of 1990 and developing countries undergoing industrialisation have been unable to agree a legal binding agreement to follow that reached for Kyoto when it expires in 2015 (Kyoto Protocol 1998). Securing such a treaty internationally is crucial if the risks associated with extreme variations in climatic conditions are to be mitigated and/or avoided before it is too late (IPCC 2014). Already, Sri Lanka is experiencing more extensive flooding and drought than in the past, and the government is seeking expert-driven technological engineering solutions to the problem instead of involving local communities in co-producing more sustainable answers for the long-term (Gunatilake 2010).

Transcending scientific-management paradigms requires the study of how actors in disaster scenarios make sense of risk, vulnerability and disaster through their everyday life practices and how these compete with rival scientific narratives to ensure that particular local stories gain the status of truth (Hilhorst 2003). Valuing local residents' inputs requires community outsiders including government officials to ask questions about how knowledge and power differentials between various partners evolve and to recognise that partnerships may limit relationships with other groups, and sometimes create or widen existing gaps between different actors. However, collaborative partnerships can seek complementarity and win-win situations that reduce people's vulnerability to disasters. The IIPP Project revealed that while there was some evidence backing Hancock's (1996) and Hoogvelt's (2007) pessimism about aid being delivered inappropriately by some donor organisations, there were many others who did not. These included the organisations involved in the IM and PPM. Additionally, the IIPP ethnographic study demonstrated that Hilhorst's (2003) three social domains for analysing disasters applied in the 2004 Indian Ocean Tsunami in Sri Lanka, these domains co-existed simultaneously and the local villagers were selective about which ones they deemed appropriate at any given point in time. Thus, they retained agency, whatever external donors thought, and this is an important lesson for anyone thinking of becoming involved in delivering humanitarian aid. I now consider the implications of responses to the more recent tragedy of Typhoon Haiyan in the Philippines.

The Philippines and Typhoon Haiyan/Yolanda

One of the strongest tropical cyclones to hit land in the Philippines was Typhoon Haiyan or Typhoon Yolanda as it was known locally in the Philippines. It reached sustained wind speeds of up to 195 mph, affected 11 million people, made 1.9 million homeless and killed 6,300 Filipinos when it struck the central Philippines on November 7, 2013. Tacloban, a large city in Leyte Island faced enormous

devastation including the death of 6,000 inhabitants, the destruction of the terminal building of its airport and the flooding of the first floor of the evacuation shelter in its convention centre when a 17-foot storm surge hit it during the Typhoon's onslaught. Damage to transportation, communication, power and medical infrastructures meant that help was seriously hampered and did not reach some of the remote villages affected for weeks (UNICEF 2013).

Desperation drove some people to take forcibly food and other supplies from warehouses in a number of sites where it was being stockpiled. This scenario depicts how social divisions, poverty, marginality and exposure complicate natural hazards to intensify vulnerability and create subsidiary disasters (Bankoff 1997; Bankoff 2002; Bankoff et al. 2013). It also indicates that people will exercise personal agency to survive, and although it would require legislation to follow through on the idea that 'taking food, water and medicines in an emergency' should not be considered an illegal act in disaster circumstances. Many jurisdictions consider 'looting' in the context of a natural disaster a serious offense with extensive penalties, e.g. in Australia, this can be punished with up to 10 years of imprisonment. Scientific rationality in relation to offending insists that such punishment is appropriate and can deter others similarly inclined. On the ground in Tacloban, practitioners sought to provide emergency water, food, clothing and shelter, despite the many challenges given the devastated built infrastructures.

Other responses were also available. As Typhoon Haiyan ravaged the Philippines, Filipino representative, Yeb Saño, began a hunger strike in sympathy with the suffering of his fellow citizens during the UNFCCC COP19 (United Nations Framework Convention on Climate Change, Conference of the Parties, 19th meeting) discussions in Warsaw, Poland in December 2013 in a dramatic move aimed at highlighting the plight of low-income countries in climate change debates that seemed to lack the political will to reach an equitable and legally binding agreement. Although others joined this initiative, the failure of their actions to achieve his objective indicate that social change is not easily forthcoming in situations in which technocratic, scientific rationalities dominate the prevailing discourses, as suggested by Hilhorst (2003). This lack of agreement holds despite the holding of the subsequent UNFCCC COP20 talks in Lima, Peru in December 2014 while Typhoon Hagupit was ripping through the Philippines.

During Hurricane Haiyan, aid from 41 countries reached the Philippines, with the key donors being the United Kingdom with US$131 million, the United States with US$86.7 million, Japan with US$52 million, Norway with US$41.6 million and Canada $40 million (Umbao 2013). Humanitarian organisations including UN agencies and NGOs worked together through a 'cluster system' to deliver aid such as food, shelter and medicines, restore health services and initiate economic activities in devastated areas. The 'cluster system' aims to coordinate activities through a specified division of labour and avoid duplication of efforts. Coordination has to occur within and between clusters, making effective communication between them vital. Clusters meet several times a year and there are subcommittees established to plan and prepare for crisis responses. Close collaboration between military and civilian personnel contribute to response effectiveness.

Cluster coordination relies on Hilhorst's (2003) disaster-governance approach to disasters and is very much a top-down model.

The cluster system was first used in 2005 following that year's earthquake in Pakistan. In the Philippines, during the aftermath of Typhoon Haiyan, the World Health Organisation (WHO) drove this process for donations linked to health care equipment and medical supplies and the provision of psychosocial support to deal with the trauma experienced by significant numbers of people. The World Food Programme led the response around food security and logistics, while UNDP (UN Development Programme) was responsible for leading on recovery and livelihoods. The International Federation of the Red Cross and Red Crescent Societies and the UN refugee agency undertook the lead for shelter. In the Haiyan disaster, Save the Children and Medair were also included in this cluster.

The decision to activate the 'cluster system' is taken by OCHA (the UN Office for the Coordination of Humanitarian Affairs) in discussion with the government of the country concerned. Each cluster works closely with the relevant ministry in the nation affected. Thus, this system relies heavily on government competence and efficiency, the absence of which can seriously hamper relief efforts as occurred in the 2010 Haiti earthquake. Following Typhoon Haiyan, the immediate response in Tacloban, the Philippines, was hampered by political rivalry between the Mayor of Tacloban and the President of the Philippines, but a public outcry led to the successful deployment of national troops for the aid endeavour a week later. This response highlights a weakness in the disaster-governance model: its top-down approach relies on having an efficient governance system capable of taking decisions as and when necessary (Hilhorst 2003).

A year later, the lessons learnt from Haiyan were put to good use during Typhoon Hagupit, known locally as Ruby, which struck the Philippines on December 7, 2014. Compared to Haiyan, which was a Category 5 storm, Hagupit was a lesser Category 3 one (UNICEF 2014). Although 1 million people were affected by Hagupit, three people died because the early warning system was effective, calls for evacuation were heeded and military personnel helped restore transportation links quickly (personal communication). People were evacuated quickly, well before Hagupit struck. Thus, lessons from Haiyan were applied to the benefit of the local population. This response demonstrated that the disaster-governance paradigm can deliver aid effectively when it is well-coordinated and functioning as a well-oiled system.

However, the Philippines' disaster-governance and scientific-management approaches to disasters face a complicating issue that they cannot transcend by operating within those paradigms. That is the thorny matter of debt repayments. Filipino society cannot ignore the considerable structural barriers that these impose in the recovery and reconstruction processes after a disaster. The obstacles this creates have the effect of worsening the country's financial woes.

Soon after Haiyan struck the Philippines, the World Bank and Asian Development Bank pledged US$1.9 billion for emergency assistance and reconstruction. While this was welcome money, receiving it as loans rather than grants added to the debt burden, making it one that the country cannot repay in the foreseeable

future. The government of the Philippines had received US$115bn in external loans and repaid US$132bn which included principal sums and interest over the past 40 years. However, it still owes US$60bn. The government was already spending 20 per cent or US$8.4 billion of its income on debt repayment in 2013, before Typhoon Haiyan devastated its terrain and peoples (Jones 2013). The Freedom from Debt Coalition is an organisation calling for the cancellation of this debt and demanding reparations for loss and damages caused by climate change. Climate change is being highlighted as a contributory factor to the large number of 'natural' disasters that hit the Philippines yearly. NGOs are calling for reparations for damage and loss caused by climate change, but international discussions on this point are stalled for the time being. Yet, the Philippines continue to be a disaster-prone country, with Typhoon Haiyan representing the 25th hurricane of 2013. It needs a solution to the deterioration in its natural resilience that climate change is aggravating. The failure of this to materialise highlights the shortcomings in both the scientific-management and disaster-governance models in dealing with disasters. Scientific arguments and well-administered solutions cannot persuade those who lack the political will to respond in ways that enhance the public good for the largest numbers.

NGOs have, through their practitioners and volunteers, played a key role in preparing citizens in the Philippines for potential disasters. Governance issues and according victim-survivors human agency are two major considerations in their interventions. The Philippine Citizen's Disaster Preparedness Centre (CDPC) is an NGO that has developed community-based disaster responses in vulnerable communities. The CDPC adopts a participatory approach that involves community residents in identifying the hazards and vulnerabilities that threaten them to support them in co-devising community-based disaster preparedness and mitigation plans. Thus, it operates according to Hilhorst's (2003) local production paradigm that seeks co-produced responses to problems. The Philippines, as Sri Lanka, have all three of Hilhorst's (2003) disaster performance paradigms operating simultaneously, but different actors seem to dominate in each type.

Conclusions

Responding to people's needs in a disaster is complex, sensitive and demanding work that constantly interrogates and crisscrosses both the local and the global. Hilhorst's (2003) three performance paradigms – the scientific-management, the disaster-governance and the local production – are all relevant in learning lessons about what approaches will yield results in different circumstances. Each paradigm has its strengths, weaknesses and purposes. The scientific-management approach is useful in understanding hazards and addressing some of the problems that these present victim-survivors so that they can be better prepared to understand the risk a particular hazard poses for their community and take steps to mitigate it. While this model overlaps to some extent with the disaster-governance one, the latter is better equipped to deal with the engineering and technical interface between people and natural hazards, and can deliver desired goods and services efficiently

and effectively, e.g. flood-proof housing, early warning systems and life sustaining materials such as debris-moving equipment, water food, clothing, medicines and shelter in the immediate aftermath of a disaster. The local production model is better at recognising human agency and in engaging residents in devising solutions that they can endorse and feel they own in responding to their needs throughout all stages of the disaster cycle if they have social capital, networks and resources to call upon. These are important in developing self-sufficiency to survive disasters and in rebuilding communities after a calamity. However, local initiatives work more effectively and are of longer duration if they combine local knowledge, skills and resources with expert (often external) ones to coproduce community-based disaster responses (Dominelli 2012, 2014).

However, a key weakness of all of Hilhorst's (2003) three paradigms is that of shifting policies in more resilient, responsive and community-empowering directions. Relying on persuading politicians who are interested in either their own or party-political national interests carries little purchase in addressing the complexities of responding to disasters. The current failure of the climate change talks to secure a legally binding, equitable agreement that reduces greenhouse gas emissions across the planet is testament to this. It is easier to convince practitioners to work across borders and boundaries, whether economic, political, cultural, social or psychological, to innovate, experiment and improvise while carrying community residents with them.

At the academic research level, the IIPP, BIOPICC and EwF projects have demonstrated the fruitfulness of interdisciplinary approaches in developing better science models and enhanced practice models. However, another issue that the past has revealed is that the interplay between the local and the global is more complex than the models currently to hand can address effectively. Thus, a key challenge for researchers, practitioners, policymakers and victim-survivors of disasters is that of co-producing new paradigms for theory and practice that can address the fluidity and multidimensionality of human agency and the multi-level responses essential to forming and retaining resilience throughout the disaster cycle from prevention to preparation, from immediate relief and recovery to post-disaster reconstruction. In this way, the knowledge, skills and resources necessary to deal with disasters in a sustainable way and one that does not jeopardise the earth's ability to meet the current and future needs of the earth's people, flora, fauna and physical resources.

If disaster work engaging communities in coproduction is done well, it can be emotionally draining and richly rewarding at the same time. A social work educator in the IIPP model articulated the issue thus:

> if any social worker or social work organization wants to really help or implement their social work intervention they should first go to the village level, live with the people, understand their needs and then make that coordination with the agencies. Then start the work from the village level *not from what the organizations or the other people want them to do* [emphasis mine]. . . . We have to understand that Sri Lanka is a culturally diverse society, the ethnic diversity is very high. The religious, we have people from four or five religions so it's a mixed

> society. When you work with people you really have to have that individual understanding of the individual person's as well as the community's needs. . . . They [social workers] have to understand this diversity. To understand that, they have to be in the village, they have to start from the village level. They have to be equipped with all these good practices in social work. They have to have a good understanding of ethics because some organizations just try to take maximum advantage of the situation. So they should have this basic understanding and they should respect, they should learn to respect, the diversity of people.
>
> (Social work educator interviewed, IIPP Project)

Engaging communities through egalitarian partnerships under their control, therefore, is one of the most important lessons past responses can teach current humanitarian workers.

Note

1 The three projects are:

The ESRC (Economic and Social Research Council) funded a three-year project entitled *Internationalising Institutional and Professional Practices* (IIPP) which contains 368 interviews, 35 sets of field notes, 10 recorded focus group discussion involving 12 Sri Lankan villages and an online survey, in evaluating humanitarian responses to the 2004 Indian Ocean Tsunami in Sri Lanka and devising a humanitarian aid toolkit and ethical guidelines for research in disaster situations;

The ESPSRC (Engineering and Physical Sciences Research Council) funded the three-year project on *Built Infrastructures, Older People and Service Provisions under Conditions of Climate Change* (BIOPICCC) which explored local responses to heatwaves and cold snaps in selected case-study sites in the UK through an interdisciplinary team that worked with older people, practitioners and policymakers to devise a toolkit that would enhance policy and practice; and

The *Earthquakes without Frontiers*, a five-year project funded by NERC (Natural Environment Research Council) which focuses on continental earthquakes, especially in China, Nepal, Kazakhstan and India and involves an extensive network of academics, practitioners and policymakers in discovering new information about seismic and associated hazards in these areas to improve disaster practice responses and policies at local, national and international levels.

References

Action Aid. (n.d.). *Real Aid: Ending Aid Dependency*. Action Aid, London.

Aird, S. (n.d.). China's Three Gorges: The Impact of Dam Construction on Emerging Human Rights. https://www.wcl.american.edu/hrbrief/08/2aird.pdf (Accessed: 21 January 2015).

Alexander, D. (1997). The study of natural disasters, 1977–1997: Some reflections on a changing field of knowledge. *Disasters*, 21(4), 284–304.

Arce, A. and Long, N. (2000) Reconfiguring modernity and development from an anthropological perspective. In: Arce, A and Long, N. (Eds.) *Anthropology, Development and*

Modernities: Exploring Discourse, Counter-Tendencies and Violence. Routledge, London. 1–31.

Bankoff, G. (2002). *Cultures of Disaster: Society and Natural Hazard in the Philippines*. Routledge, London and Curzon, New York.

Bankoff, G., Ferks, G. and Hilhorst, D. (2013). *Mapping Vulnerability: Disasters, Development and People*. Routledge, London.

Bennett, R. and Daniel, M. (2002). Media reporting of third world disasters: The journalist's perspective. *Disaster Prevention and Management: An International Journal*, 11(1), 33–42.

Blaikie, P., Cannon, T., Davis, I. and Wisner, B. (1994). *At Risk: Natural Hazards, People's Vulnerability, and Disasters*. Routledge, London.

Bookstein, A. (2003). *Beyond the Headlines: An Agenda for Action to Protect Civilians in Neglected Conflicts*. Oxfam Publishing, Oxford.

Casadevall, T. (1993). Volcanic hazards and aviation safety: Lessons from the past. *Flight Safety Digest*. Retrieved May. pp. 1–9. www.caa.co.uk/docs/375 (Accessed: 13 March 2010).

Cooper, R. (2014). Slumdog millionaire district where doctors and lawyers live side by side with the poor. *Daily Mail*. Retrieved 11 January. http://www.dailymail.co.uk/news/article-2526887/Slumdog-Millionaire-district-doctors-lawyers-live-Squalid-poverty-stricken-Mumbai-no-one-wants-incredible-community-spirit.html (Accessed: 12 January 2015).

Crutzen, P. and Schwägeri, C. (2011). Living in the Anthropocene: Toward a New Global Ethos. http://e360.yale.edu/feature/living_in_the_anthropocene_toward_a_new_global_ethos/2363/ (Accessed: 20 December 2014).

Dominelli, L. (2004) *Social Work: Theory and Practice for a Changing Profession*. Polity Press, Cambridge.

Dominelli, L. (2009) *Social Work in a Globalising World*. Polity Press, Cambridge

Dominelli, L. (2012). *Green Social Work*. Polity Press, Cambridge.

Dominelli, L. (2013a). Gendering climate change: Implications for debates, policies and practices. In Alston, M. and Whittenbury, K. (eds.), *Research, Action and Policy: Addressing the Gendered Impacts of Climate Change*. Springer, Sydney. 77–94.

Dominelli, L. (2013b). Mind the Gap: Built infrastructures, sustainable caring relations and resilient communities in extreme weather events. *Australian Social Work*, 66(2), 204–217.

Dominelli, L. (2014). The opportunities and challenges of social work interventions in disaster situations. *International Social Work*, 57(4), 337–334.

Douglas, M. and Wildavsky, A. (1982). *Risk and Culture: An Essay on the Selection of Technological and Environmental Dangers*. University of California Press, Berkeley.

Duffield, M. (1993) Disaster relief and asset transfer in the horn: Political survival in a permanent emergency. *Development and Change*, 24, 131–137.

Enarson, E. and Morrow, B. (1998). *The Gendered Terrain of Disaster: Through Women's Eyes*. Florida International University, Miami.

Gunatilake, M. (2010). International Aid and Surviving Drought in Sri Lanka. *Seminar at the institute of hazard, risk and resilience*. Durham University, Durham, June.

Hancock, G. (1996). *Lords of Poverty: The Power, Prestige, and Corruption of the International Aid Business*. Atlantic Monthly Press, New York.

Harvey, D. (1990). *The Condition of Postmodernity: An Inquiry into the Conditions of Cultural Change*. Blackwell Publishers, Oxford.

Hilhorst, D. (2003). Responding to disasters: Diversity of bureaucrats, technocrats and local people. *International Journal of Mass Emergencies and Disasters*, 21(1), 37–55.

Hirst, P., Thompson, G., and Bromley, S. (2009) *Globalization in Question*. Polity Press, Cambridge.

Hoogvelt, A. (2007). Globalisation and imperialism: Wars and humanitarian intervention. In Dominelli, L. (ed.), *Revitalising Communities in a Globalising World*. Ashgate, Aldershot. 7–20.

IPCC (Intergovernmental Panel on Climate Change). (2014). Fifth Assessment Report. http://www.ipcc.ch/report/ar5/ (Accessed: 20 December 2014).

Jansen, J. (1998) *But Our Natives Are Different: Race, Knowledge and Power in the Academy*, Pretoria University, Pretoria. http://repository.up.ac.za/bitstream/handle/2263/2008/Jansen%20%281998%29c.pdf?sequence=1&isAllowed=y (Accessed: 23 October 2012).

Jones, T. (2013). The Philippines Is Devastated as Much by Unfair Debt as Typhoon Haiyan. http://www.theguardian.com/global-development/poverty-matters/2013/dec/23/philippines-devastated-unfair-debt-typhoon-haiyan (Accessed: 12 January 2014).

Klein, N. (2008). *The Shock Doctrine*. Henry Holt and Company Inc, New York.

Kyoto Protocol. (1998). http://unfccc.int/resource/docs/convkp/kpeng.pdf (Accessed: 13 December 2013).

Lane, S., Odoni, N., Landström, C., Whatmore, S. J., Ward, N. and Bradley, S. (2011). Doing flood risk science differently: An experiment in radical scientific method. *Transactions*, 36, 15–36.

Larsen, G. (1999). Gosið í Eyjafjallajökli 1821–1823 [The eruption of the Eyjafjallajökull volcano in 1821–1823]. Research Report RH-28–99. Science Institute, Reykjavík.

Lay, T., Kanamori, H., Ammon, C., Nettles, M., Ward, S., Aster, R., . . . Sipkin, S. (2005). The great sumatra-andaman earthquake of 26 December 2004. *Science*, 308, 1127–1133.

Manyena, B. (2006). The concept of resilience revisited. *Disasters*, 30(4), 434–450.

McGuire, B. (2014). Waking the Giant: How a Changing Climate Triggers Earthquakes, Tsunamis and Volcanoes. http://e360.yale.edu/feature/could_a_changing_climate_set_off_volcanoes_and_quakes/2525/ (Accessed: 20 December 2014).

Medford, W. and Dominelli, L. (2013). Non-Governmental Organisations Involvement in Post-Tsunami Responses in Sri Lanka: Insights from an On-Line Survey. Durham University, SASS unpublished paper, Durham.

Osterhammel, J. (2005). *Colonialism: A Theoretical Overview*. Markus Wiener Publishers, Princeton, NJ.

Oven, K., Curtis, S. E., Reaney, S., Riva, M., Stewart, M., Ohlemüller, R., . . . Holden, R. (2012). Climate change and health and social care: Defining future hazard, vulnerability and risk for infrastructure systems supporting older people's health care in England. *Applied Geography*, 33(2), 16–24.

Pelling, M. and Uitto, J. (2001). Small island developing states: Natural disaster vulnerability and global change. *Environmental Hazards*, 3, 49–62.

Pilger, J. (2005). The Other Tsunami: Cover Story in New Statesman. 10 January. http://www.johnpilger.com (Accessed: 20 January 2005).

Pittaway, E., Bartolomei, L. and Rees, S. (2007). Gendered dimensions of the 2004 tsunami and a potential social work response in post-disaster situations. *International Social Work*, 50(3), 307–319.

Possekel, A. (1999). *Living with the Unexpected: Linking Disaster Recovery to Sustainable Development in Montserrat*. Springer-Verlag, Berlin/Heidelberg/New York/Tokyo.

Rijkswaterstaat. (2002). *Ruimte voor de Rivier. Startnotitie MER in het kader van PKB Procedure*. Rijkswaterstaat, Den Haag.

Umbao, E. (2013). International Donations to the Philippines for Yolanda Victims. http://philnews.ph/2013/11/12/international-donations-philippines-typhoon-yolanda-victims/ (Accessed: 21 January 2014).

UNICEF. (2013). Typhoon Haiyan. http://www.unicefusa.org/mission/emergencies/hurricanes/2013-philippines-typhoon-haiyan (Accessed: 8 January 2014).

UNICEF. (2014). Typhoon Haigput Hits the Philippines. http://www.unicefusa.org/stories/typhoon-hagupit-ruby-hits-philippines/20766 (Accessed: 30 December 2014).

Yip, K. Y. (2005) A dynamic Asian response to globalization in cross-cultural social work. *International Social Work*, 48(5), 593–607.

Index